Insect Nicotinic Acetylcholine Receptors

ADVANCES IN EXPERIMENTAL MEDICINE AND BIOLOGY

Recent Volumes in this Series

Volume 675
RECENT ADVANCES IN PHOTOTROPHIC PROKARYOTES
Edited by Patrick C. Hallenbeck

Volume 676
POLYPLOIDIZATION AND CANCER
Edited by Randy Y.C. Poon

Volume 677
PROTEINS: MEMBRANE BINDING AND PORE FORMATION
Edited by Gregor Anderluh and Jeremy Lakey

Volume 678
CHEMO FOG: CANCER CHEMOTHERAPY-RELATED COGNITIVE IMPAIRMENT
Edited by Robert B. Raffa and Ronald J. Tallarida

Volume 679
MIPS AND THEIR ROLE IN THE EXCHANGE OF METALLOIDS
Edited by Thomas P. Jahn and Gerd P. Bienert

Volume 680
ADVANCES IN COMPUTATIONAL BIOLOGY
Edited by Hamid R. Arabnia

Volume 681
MELANOCORTINS: MULTIPLE ACTIONS AND THERAPEUTIC **POTENTIAL**
Edited by Anna Catania

Volume 682
MUSCLE BIOPHYSICS: FROM MOLECULES TO CELLS
Edited by Dilson E. Rassier

Volume 683
INSECT NICOTINIC ACETYLCHOLINE RECEPTORS
Edited by Steeve Hervé Thany

A Continuation Order Plan is available for this series. A continuation order will bring delivery of each new volume immediately upon publication. Volumes are billed only upon actual shipment. For further information please contact the publisher.

Insect Nicotinic Acetylcholine Receptors

Edited by

Steeve Hervé Thany, PhD

*Laboratoire Récepteurs et Canaux Ioniques Membranaires, RCIM
UPRES EA 2647/USC INRA 2023, IFR 149 QUASAV
Université d'Angers, UFR de Sciences, Angers, France*

Springer Science+Business Media, LLC

Landes Bioscience

Springer Science+Business Media, LLC
Landes Bioscience

Printed in the USA.

Springer Science+Business Media, LLC, 233 Spring Street, New York, New York 10013, USA
http://www.springer.com

Please address all inquiries to the publishers:
Landes Bioscience, 1002 West Avenue, Austin, Texas 78701, USA
Phone: 512/ 637 6050; FAX: 512/ 637 6079
http://www.landesbioscience.com

The chapters in this book are available in the Madame Curie Bioscience Database.
http://www.landesbioscience.com/curie

Insect Nicotinic Acetylcholine Receptors, edited by Steeve Hervé Thany. Landes Bioscience / Springer Science+Business Media, LLC dual imprint / Springer series: Advances in Experimental Medicine and Biology.

ISBN: 978-1-4419-6444-1

Library of Congress Cataloging-in-Publication Data

Insect nicotinic acetylcholine receptors / edited by Steeve Hervé Thany.
p. ; cm. -- (Advances in experimental medicine and biology ; v. 683)
Includes bibliographical references and index.
ISBN 978-1-4419-6444-1
1. Nicotinic receptors. 2. Insecticides--Physiological effect. 3. Insects--Physiology. I. Thany, Steeve Herve, 1972- II. Series: Advances in experimental medicine and biology, v. 683. 0065-2598 ;
[DNLM: 1. Receptors, Nicotinic. 2. Insecticides--toxicity. 3. Insects--physiology. 4. Synaptic Transmission. W1 AD559 v. 683 2010 / WL 102.8 I59 2010]
QP364.7.I565 2010
632'.9517--dc22

2010011799

DEDICATION

To my son, Lucien

PREFACE

The aim of this volume is to summarize our understanding on the insect nicotinic acetylcholine receptors. This area of research received great impetus from the identification of the first subunit sequences to be used as neonicotinoid insecticide target sites. The first chapter illustrates the finding that the insect central nervous system is extremely rich in acetylcholine receptors that have a predominantly nicotinic pharmacology. Chapter 2 shows that these receptors have an overall structure that is reminiscent of the Cys-loop family of ligand-gated ion channels. In Chapter 3, analysis of genome sequences has shown that nAChR gene families remain compact in diverse insect species, when compared to their nematode and vertebrate counterparts, containing 10-12 nicotinic acetylcholine receptor genes. Chapter 4 demonstrates that several amino acids that account for ligand binding domain are conserved in the insect nicotinic receptor subunit. Pharmacological properties of native insect nicotinic receptors using electrophysiological studies are described in Chapter 5, while Chapter 6 indicates the types of strategies being used by electrophysiologists to study the subunit composition of nicotinic receptor using hybrid receptors. Chapters 7 and 8 provide basic information on neonicotinoid insecticides, the most important new class of synthetic insecticides of the past three decades and their toxicity on the honeybee colonies. Chapter 9 explores the involvement of insect nicotinic receptors in learning and memory processes using the honeybee as insect model.

Although a book of this nature can provide the details only of commonly published results, it is hoped that it may provide a useful guide to the newcomer to the field as well as to point out some of the future challenges. For example, we need to determine the precise subunit nomenclature of insect nicotinic receptors. This nomenclature varies amongst species and this led to some of the early confusion that persists. We need to be precise in identifying the subunit composition of native insect nicotinic receptor subtypes, their functional properties and physiological role.

Steeve Hervé Thany, PhD
Laboratoire Récepteurs et Canaux Ioniques Membranaires, RCIM
UPRES EA 2647/USC INRA 2023, IFR 149 QUASAV
Université d'Angers, UFR de Sciences, Angers, France

ABOUT THE EDITOR...

STEEVE HERVÉ THANY is an assistant Professor for Neurobiology and Neurophysiology at the University of Angers, France. His research interests are in the areas of molecular biology, electrophysiology, pharmacology and behavioral processes involving insect nicotinic acetylcholine receptors. Recent research focused on the effects of neonicotinoid insecticides on nicotinic receptors and the intracellular mechanisms regulating their function. He is a member of numerous national and international scientific organisations, including the French Society for Neuroscience (SN) and the Federation of European Neuroscience Societies (FENS).

PARTICIPANTS

Axel Decourtye
Association de Coordination Technique Agricole, ACTA
Institut Claude Bougelat
Marcy L'Etoile
France

Joseph A. Dent
Department of Biology
McGill University
Montreal, Quebec
Canada

James Devillers
Centre de Traitement de l'Information Scientifique, CTIS
Rillieux La Pape
France

Monique Gauthier
Centre de Recherches sur la Cognition Animale, CRCA
UMR 5169 CNRS UPS
Toulouse
France

Andrew K. Jones
MRC Functional Genomics Unit
Department of Physiology, Anatomy and Genetics
University of Oxford
Oxford
UK

Stuart J. Lansdell
Department of Neuroscience, Physiology and Pharmacology
University College London
London
UK

Bruno Lapied
Laboratoire Récepteurs et Canaux Ioniques Membranaires, RCIM
UPRES EA 2647/USC INRA 2023, IFR 149 QUASAV
Université d'Angers, UFR de Sciences
Angers
France

Neil S. Millar
Department of Neuroscience, Physiology and Pharmacology
University College London
London
UK

David B. Sattelle
MRC Functional Genomics Unit
Department of Physiology, Anatomy and Genetics
University of Oxford
Oxford
UK

Steeve Hervé Thany
Laboratoire Récepteurs et Canaux Ioniques Membranaires, RCIM
UPRES EA 2647/USC INRA 2023, IFR 149 QUASAV
Université d'Angers, UFR de Sciences
Angers
France

Hélène Tricoire-Leignel
Laboratoire Récepteurs et Canaux Ioniques Membranaires, RCIM
UPRES EA 2647/USC INRA 2023, IFR 149 QUASAV
Université d'Angers, UFR de Sciences
Angers
France

CONTENTS

ABBREVIATIONS

Ac-AChBP	AChBP from *Aplysia californica*
ACh	Acetylcholine receptor
AChBP	Acetylcholine binding protein
AChE	Acetylcholinesterase
AcSCh	Acetylthiocholine
ADAR	Adenosine deaminases acting on RNA
AOX	Cytosolic aldehyde oxidase
α-BGT	Alpha-bungarotoxin
α-CBX	Alpha-cobratoxin
ATP	Adenosine triphosphate
AL	Antennal lobe
BuCh	Butyrylcholine
BuChE	Butyrylcholinesterase
BuSCh	Butyrylthiocholine
CaMKII	Calcium/calmodulin-dependent protein kinase
cAMP	Cyclic adenosine monophosphate
ChAT	Choline acetyltransferase
CNS	Central nervous system
ColQ	Collagenous 'tail'
CR	Conditioned response
CS	Conditioned stimulus
DAG	Diacylglycerol
DUM	Dorsal unpaired median
DβHE	Dihydroxy-β-erythroidine
DiC8	Diacylglycerol analogue
d-TC	d-tubocurarine
EPI	Epibatidine
G-protein	Guanine nucleotide binding protein
GDP	Guanosine diphosphate
GDP-βS	Guanosine -5'-O(2-thiodiphophate)
GPI	Glycophosphatidylinositol

GTP	Guanosine triphosphate
GTPγS	Guanosine 5'-O(3-thiotriphosphate)
G_i	Guanine nucleotide binding protein which regulates inhibition of adenylate cyclase
G_o	'other' G-protein
IMI	Imidacloprid
IP	Inositol phosphate
IP_2	Inositol biphosphate
IP_3	Inositol triphosphate
IP_4	Inositol tetrakisphosphate
KC	Kenyon cell
lATG	Lateral antennoglomerular tract
LD50	Median lethal dose
Ls-AChBP	AChBP from *Lymnea stagnalis*
LTM	Long term memory
mATG	Median antennoglomerular tract
MB	Mushroom bodies
MEC	Mecamylamine
MLA	Methyllycaconitine
MTM	Median-term memory
nAChRs	Nicotinic acetylcholine receptors
nAChD	Desensitized nicotinic acetylcholine receptor
nAChN	Non-desensitized nicotinic acetylcholine receptor
OA	Octopamine
PER	Proboscis extension reflex
PIP	Phosphatidylinositol 5-monophosphate
PIP2	Phosphatidylinositol 4,5-bisphosphate
PKA	Protein kinase A
PKC	Protein kinase C
PLC	Phospholipase C
pLGIC	Ligand-gated ion channel
PrCh	Propionylcholine
PrSCh	Propionylthiocholine
RIC-3	Gene resistance to inhibitors of cholinesterase
SCAM	Substituted-cysteine-accessibility method
SOG	Suboesophageal ganglia
TM	Transmembrane domain
US	Unconditioned stimulus
VUM	Ventral unpaired median
5-HT	5-hydroxytryptamine

ACKNOWLEDGEMENTS

Editor thanks all contributors for the time spent in the publication of this book.

CHAPTER 1

Identification of Cholinergic Synaptic Transmission in the Insect Nervous System

Steeve Hervé Thany,* Hélène Tricoire-Leignel and Bruno Lapied

Abstract

A major criteria initially used to localize cholinergic neuronal elements in nervous systems tissues that involve acetylcholine (ACh) as neurotransmitter is mainly based on immunochemical studies using choline acetyltransferase (ChAT), an enzyme which catalyzes ACh biosynthesis and the ACh degradative enzyme named acetylcholinesterase (AChE). Immunochemical studies using anti-ChAT monoclonal antibody have allowed the identification of neuronal processes and few types of cell somata that contain ChAT protein. In situ hybridization using cRNA probes to ChAT or AChE messenger RNA have brought new approaches to further identify cell bodies transcribing the ChAT or AChE genes. Combined application of all these techniques reveals a widespread expression of ChAT and AChE activities in the insect central nervous system and peripheral sensory neurons which implicates ACh as a key neurotransmitter.

The discovery of the snake toxin alpha-bungatoxin has helped to identify nicotinic acetylcholine receptors (nAChRs). In fact, nicotine when applied to insect neurons, resulted in the generation of an inward current through the activation of nicotinic receptors which were blocked by alpha-bungarotoxin. Thus, insect nAChRs have been divided into two categories, sensitive and insensitive to this snake toxin. Up to now, the recent characterization and distribution pattern of insect nAChR subunits and the biochemical evidence that the insect central nervous system contains different classes of cholinergic receptors indicated that ACh is involved in several sensory pathways.

Introduction

The cholinergic pathway is one of the most important excitatory neurotransmission system in insect nervous system. The neurotransmitter; acetylcholine (ACh), synthesized by cholinergic neurons, is widely distributed throughout the nervous system and plays a key role in the insect synaptic neurotransmission (Fig. 1).[1,2] Two major types of ACh receptors have been characterized in insects: muscarinic ACh receptors (mAChRs) and nicotinic ACh receptors (nAChRs). Numerous studies have revealed the pharmacological properties of mAChRs and nAChRs on

*Corresponding Author: Steeve Hervé Thany—Laboratoire Récepteurs et Canaux Ioniques Membranaires (RCIM), UPRES EA 2647/USC INRA 2023, IFR 149 QUASAV, Université d'Angers, UFR de Sciences. 2 Bd. Lavoisier 49045 Angers, France.
Email: steeve.thany@univangers.fr

Insect Nicotinic Acetylcholine Receptors, edited by Steeve Hervé Thany.

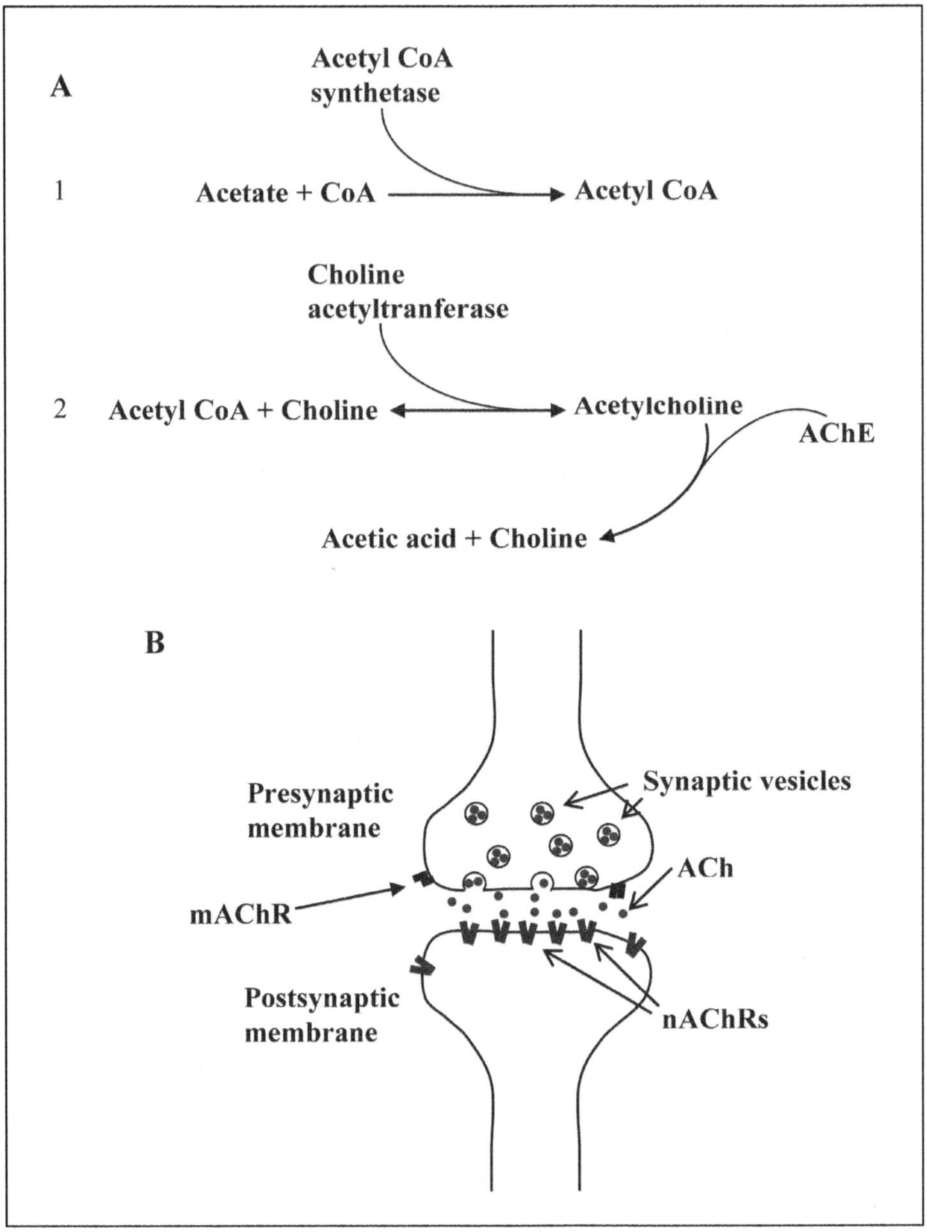

Figure 1. Schematic representation of the synthesis and hydrolysis of ACh.

synapse-free cell bodies as well as on the synapse-neuropile complex.[3-5] mAChRs have been reported as crucial element in the regulation of ACh release in insect synaptic neurotransmission. If it is well known that ACh interacts with postsynaptic nAChRs, the participation of mAChRs at presynaptic level has also been demonstrated electrophysiologically.[6,7] In this case, mAChRs sharing pharmacological properties very similar to vertebrate M2 mAChR-subtypes are involved in the modulation of ACh release from presynaptic endings.[7] It has been demonstrated, particularly in cockroach central nervous system, that any means that increase ACh concentration within the synaptic cleft result in direct activation of presynaptic mAChRs involved in the negative feedback mechanism. At postsynaptic level, nAChRs are of particular interest because it has been shown that they are (1) involved in fast excitatory synaptic transmission and (2) are target sites for insecticides, particularly neonicotinoid insecticides.[8-10]

Additional complementary approaches including histochemical and immunocytochemical detection, electrophysiology and autoradiographic studies of choline uptake[11-15] have characterized the different physiological component of the cholinergic system corresponding to the ACh-synthesizing and –degrading enzymes such as choline acetyltransferase (ChAT) and acetylcholinesterase (AChE), respectively. Their physiological implications have been clearly demonstrated since null mutations in either the ChAT or AChE genes result in the late embryonic lethality indicating the essential nature of ACh pathways in the insect.[16]

Because ChAT and AChE activities confirm the existence of both types of cholinergic receptor (i.e., mAChR and nAChR), alpha-bungarotoxin (α-Bgt)-binding studies have also been used to further identify, pharmacologically, insect nAChRs as sensitive- and insensitive-nAChR.[17] α-Bgt is the most useful tool in biochemical and pharmacological studies of putative nAChRs. This toxin is a component of the venom of the snake *Bungarus multicinctus*, which is a highly potent antagonist of nAChRs. Furthermore, it has been demonstrated that the insect nervous system contains high concentrations of α-Bgt binding sites.[18-20] Because α-Bgt blocks ACh-evocked excitatory postsynaptic potential (EPSP) in insect ganglia[3,21] the central role of nAChR subtypes in cholinergic synaptic transmission have been clearly established.[17,22] In the following sections we will discuss on the identification of cholinergic neurons and their distribution in the insect central nervous system (CNS).

Insect Acetylcholinesterase: Catalytic Properties and Tissue Distribution

Generally, it is suggested that the enzymes involved in ACh biosynthesis, ChAT and Acetyl CoA synthetase, are concentrated in the cytoplasm of cholinergic nerve terminals since ACh is thought to be synthesized mostly in these terminals. Conjugated choline is an indirect assay to identify cholinergic neurons because it is a metabolic product of ACh hydrolysis by AChE. The mode of ACh/AChE interaction is a matter of considerable interests. Generally, AChE is a type B carboxylesterase that rapidly hydrolyzes ACh at cholinergic synapses but also neuromuscular junctions. The AChE active sites are composed of an esteric region containing a nucleophilic serine residue that reacts covalently with acyl esters and an acid group interacting with ether the oxygen of ACh by hydrogen binding and an anionic site that interacts electrostatically with the onium head.[23,24] Hydrolysis of ACh involves an acetylation-deacetylation cycle. It has previously been demonstrated that the rate-limiting step is the deacetylation of the enzyme in vertebrates whereas it is the acetylation step in insect.[25] AChE occurs in several structurally distinct forms that can be differentiated by their number and type of subunits. Usually, in vertebrates, there are two acetylcholinesterases named AChE and butyrylcholinesterase (BuChE), both characterized by their capacity to hydrolyze very rapidly choline esters. Two main criteria allow to distinguish AChE from BuAChE. First, AChE and BuAChE seem to be typically blocked by different inhibitors such as carbamates and organophosphates. Second, vertebrates AChE hydrolyzes ACh or its analog acetylthiocholine (AcSCh) faster than propionylcholine (PrCh) or propionylthiocholine (PrSCh). Furthermore, it seems that AChE has not effect on butyrylcholine (BuCh) or butyrylthiocholine (BuSCh) while BuChE hydrolyzes BuCh or BuSCh.[24,26] In insects, despite the fact that AChE appeared to hydrolyze BuCh or BuSCh,[27] it is the only cholinesterase characterized today which is usually distinguished from others esters.

In vertebrates, the different forms of AChE correspond to different quaternary structures and exhibit different anchoring modes. AChE could exist as globular forms corresponding to the globular monomeric (G1), dimeric (G2) and tetrameric (G4) forms of catalytic subunits. These forms may be hydrophilic or amphiphilic. Asymmetric forms could also exist as one (A4) , two (A8) or three (A12) catalytic tetramers attached via a collagenous 'tail' (ColQ) or hydrophobic protein (P subunit) to the extracellular matrix.[28,29] In invertebrates, there are essentially globular forms. In insect, the amphiphilic form is generally prevalent. The mode of attachment of the amphiphilic forms in many insects is a glycophosphatidylinositol (GPI) anchor. This GPI anchor is covalently attached to a hydrophobic domain of the AChE-terminus. It has also been demonstrated that AChE possessed a molecular polymorphism with regard to the existence of amphiphilic and hydrophilic dimers of catalytic subunits which is present at all developmental stages.[29] Studies from *Musca* and *Drosophila* have suggested that the major form of AChE is an amphiphilic dimer which can be converted into a hydrophilic dimer and a hydrophilic monomer.[27,30,31] It is widely distributed in the brain, the thoracic and abdominal segments and the abdominal ganglia. The preferential distributions included the neuropiles of the CNS such as protocerebral and tritocerebral neuropiles, the lobula and medulla of the optic lobes, the central body, the internal of the calyces and the antennal glomeruli.[14,32-34]

Identification and Tissue Distributions of Choline Acetyltransferase in Insects

The most reliable marker of cholinergic neurons is the presence of ChAT, the enzyme catalyzing the synthesis of ACh. Cholinergic neurons can be detected by (1) using antibodies against ChAT, (2) *in situ* hybridization to detect the transcript and (3) staining for a reporter gene fused to the regulatory sequence of the ChAT gene.[35-37] In this case, ChAT has been first identified in the larval neuropil, including the larval antennal and optic lobes and in the gustatory target regions, which appeared to be due to the terminals of cholinergic afferents.[38-44] In *Drosophila melanogaster*, ChAT exists in cholinergic nerve terminals in at least two forms; a soluble and membrane-bound form that exhibit the same apparent molecular weight (75 kDa).[35] It should be noted that purification of locust ChAT results in a band at 65 kDa[45] very similar to that of reported in the spider *Cupiennius salei*.[46] The main reason that explains such difference may be that different forms of the enzyme exist.[41]

Histochemical detection of esterase activity and immunohistochemical localization of ChAT revealed that both enzymes are widely expressed in the insect nervous system. In fact, anti-ChAT immunoreactivity is well correlated to regions expressing high esterase activity.[14] In the locust ocellar visual system for instance, ChAT is also localized within synaptic clefts of L-neurons in both of the brain regions where L-neurons make excitatory and inhibitory output synapses. This suggests that ACh released by ocellar L-neurons can evoke fast excitatory and inhibitory postsynaptic potentials in different postsynaptic neurons.[40,41] Nevertheless, caution is needed in interpreting the significance of the presence of ChAT and AChE activities because staining of AChE has also been found in noncholinergic regions such as glial cells.[24,44] It seems that ACh in these cells could be involved in the axon guidance and elongation during development.[43,47]

Identification of Native Neuronal Nicotinic Acetylcholine Receptors

To obtain a first indication of the nAChR subtypes present in the insect CNS, certain α-neurotoxins such as α-Bgt was used as tool to identify specific high affinity binding sites. Thus, membrane extracts prepared from the insect CNS are rich in specific [^{125}I]-α-Bgt sites, exceeding the number of sites probed by the radiolabelled mAChR ligand quinuclidinyl benzilate.[20,48-50] Specific binding of radiolabelled α-Bgt has also been demonstrated in specific neuropiles of *Drosophila melanogaster*, *Manduca sexta*, *Acheta domesticus* and *Periplaneta americana*.[3,51-54] In fact, Schmidt-Nielsen and its colleagues have detected [^{125}I]-α-Bgt binding in the head and thorax of *Drosophila*.[51] In the honeybee brain, binding sites are confined to neuropil areas, particular to those with primary and higher order sensory projections.[15] The distribution of [^{125}I]-α-Bgt binding resembles to that reported in D. *melanogaster* and *M. sexta*. In addition, similar distribution has also been identified in the abdominal ganglia of *P. americana* and *A. domesticus* at sites of synaptic connexion.[3,55] In some insects, a single high affinity α-Bgt binding site (with similar pharmacological profile) has been identified. These results differ from recent findings showing the presence of both high and low affinity binding sites in *Myzus persicae* membranes.[56] In fact, data for [^{125}I]-α-Bgt saturable binding are consistent with two binding components, one of high affinity (K_d = 1.2 nM) and one of low affinity (K_d = 33 nM). This multiple α-Bgt binding sites suggest the presence of either multiple receptor subtypes or the cooperativity between nicotinic binding sites on the same receptor[56]. Because of these potential multiple α-Bgt sites, other studies have used additional radioligand binding studies with the norditerpenoid alkaloid methyllicaconitine (MLA) isolated from species of *Delphinium spp* (larkspurs) to confirm α-Bgt high affinity binding sites. MLA is a competitive nicotinic antagonist which is more discriminating than α-Bgt.[57] The distribution of [^{3}H]MLA and [^{125}I]-α-Bgt bindings sites have been compared in the brain of *Manduca sexta*. The patterns of labelling with both ligands are similar in the neuropil areas of the brain with a specific [^{125}I]-α-Bgt binding in an external area of the lamina and adjacent to the tapetum, a reflective tracheal structure at the base of the retina.[58] It is interesting to note that the isotopic dissociation of [^{3}H]-α-Bgt from *Myzus persicae* membranes is accelerated when initiated by MLA. These dissociation kinetics seem to be consistent with a model in which at least the majority of [^{3}H]-α-Bgt high affinity binding sites have two different ligand binding pocket (or domain) at each nicotinic receptor.[56]

Recently, neonicotinoid insecticides such as imidacloprid and azidonicotinoid have been used as ligand in order to identify specific nAChR subtypes. In *D. melanogaster*, the [^{3}H]IMI binding site is distinct from that for either [^{3}H]EPI and [^{3}H]α-Bgt. Saturable binding of [^{3}H]IMI to membranes of the peach potato aphid *Myzus persicae* revealed the presence of both high- and low-affinity [^{3}H]IMI binding sites having distinct dissociation constants and maximal binding capacities. Moreover, saturable [^{3}H]IMI binding analyzed on different insects, namely, *Nephotellix cincticeps* (Hemiptera: Cicadellidae), *Periplaneta americana* (Dictyoptera), *Lucilia sericata* (Diptera: Calliphoridae), *Drosophila melanogaster* (Canton S) (Diptera: Drosophiloidea), *Manduca sexta* (Lepidoptera: Sphingidae), *Heliothis Virescens* (Lepidoptera: Noctuidae) and *Ctenocephalides felis* (Siphonaptera: Pulicidae), have demonstrated that only aphids and green leaf hoppers (both hemiptera) have very high-affinity binding sites for [^{3}H]IMI.[59] This may indicate that only a single class of binding site is present and/or that there is no cooperativity between binding sites.

Insect Nicotinic Acetylcholine Receptors Subunit Localizations

Over the last few years, different type of nAChRs have been discriminated electro-pharmacologically and numerous nAChR subunits have been cloned from several insect species.[60-64] Their localization identified from some insect including fly *Drosophila melanogaster*,[65-68] house fly *Musca domestica*,[69] locusts *Locusta migratoria*[70] cockroach *Periplaneta americana*[21,22] and honeybee *Apis mellifera*.[63,71,72] Northern blot analysis and *in situ* hybridization from the *Drosophila* transcripts encoding ALS, Dα2, SBD and ARD subunits are detected from 10 to 12h embryos.[73-76] Similar staining of honeybee Apisα3 mRNA have been found in the late larvae around day 5 after egg laying[71] and in the *Locusta migratoria* Locα1, Locα2 and Locβ2 subunits around day 7,[70] showing that nAChR subunit expression begin in the earlier insect stage. Despite the fact that it is difficult to identify the precise stained structures, it appeared that nAChR subunit expressions are localized into specific regions. In fact, during the late embryogenesis, intense staining with an SBD probe is observed in the *D. melanogaster* sub- and supraosoephageal ganglia and the ventral cord. Hybridization with an ALS-specific probe also results in staining of the same parts of the embryonic CNS.[77] Similar expression is also detected at day 8 after egg laying, in the honeybee suboesophageal ganglion, which is found at the origin of the mandibular, maxillar and labial nerves that supply the mouthparts.[71] In the adult, immunohistochemical studies performed with anti-Dα3, anti-ARD and anti-Dα7 antibodies show a strong staining into the neuropils of the lamina, the medulla, the lobula and the lobula plate.[66,67] Similarly, the SBD and ALS subunits have also a distribution in the adult optic lobes.[65] Because ALS, Dα2 and SBD can be copurified by α-Bgt affinity chromatography, this specific colocalization allowed to consider the existence of an nAChR subtype including these subunits in the optic lobes. It is interesting to note that similar expression pattern has been identified into the honeybee *Apis mellifera* optic lobes for Apisα2, Apisα3 and Apisα7-2.[72] Regarding the subunit expressions in the adult neuropil, it is tempting to consider that nAChRs including these subunits have important function in the visual system. This has been recently studied using Dα7 subunit mutant flies. In this case, mutant flies have a defect in the jump circuit. Mutation disrupts sensory input to the giant fibers which thereby alters the *D. melanogaster* escape behavior.[67]

Conclusion

nAChRs mediate synaptic signal transmission in insect CNS. ACh release from the presynaptic nerve terminal of cholinergic synapses binds to its recognition site at postsynaptic AChR and induces conformational changes of the ionotropic receptor proteins. Both AChE and ChAT activities expressed at cholinergic synapses demonstrate the key role of ACh in the central nervous system of insects, in which ACh is a major excitatory neurotransmitter. Nevertheless, the presence of AChE and ChAT in noncholinergic tissues raises the question of a possible role of AChE distinct from the classical hydrolysis of ACh. In fact, it has previously been reported that this enzyme presents a remarkably wide range of noncholinergic functions such as cell adhesion, proliferation and neurite outgrowth.[78-80] The use of tritiated nicotinic agonists (nicotine, cytisine, methylcarbamylcholine...) or antagonists (α-bungarotoxin and methyllicaconitine) have given new insights on the characterisation of high and low affinity of nicotinic agonist binding sites. Thus nAChR subtypes have been divided in two categories, sensitive and insensitive to the α-Bgt. The hypothesis that the major neuronal nAChR is a homomeric molecular structure is advocated by the concordance between [^{125}I]-α-Bgt binding sites and presumably vertebrate homomeric α7-like receptors. But, the differences in binding studies reported in some insects suggest that they may contain several types of nAChRs, as it is the case in vertebrates. Finally, the criteria for identifying nAChR subtypes and functional properties of cholinergic pathways will be the isolation of insect nAChR subtypes. Such studies are now in progress.

References

1. Sattelle DB, David JA, Harrow ID et al. Action of alpha-bungarotoxin on identified insect central neurones. In: Sattelle DB, Hall LM, Hildebrand JG, eds. Receptors for Neurotransmitters, Hormones and Pheromones in Insects. Amsterdam: Elsevier/North Holland, Biomedical Press 1980:125-139.
2. Breer H, Sattelle DB. Molecular properties and functions of insect acetylcholine receptors. J Insect Physiol 1987; 33:771-790.
3. Sattelle DB, Harrow ID, Hue B et al. α-bungarotoxin blocks excitatory synaptic transmission between cercal sensory neurones and giant interneurone 2 of the cockroach, Periplaneta americana. J Exp Biol 1983; 107:473-489.
4. Benson JA. Electrophysiological pharmacology of the nicotinic and muscarinic cholinergic responses of isolated neuronal somata from locust thoracic ganglia. J Exp Biol 1992; 170:203-233.
5. Trimmer BA. Current excitement from insect muscarinic receptors. Trends Neurosci 1995; 18(2):104-11.

6. Hue B, Lapied B, Malecot CO. Do presynaptic muscarinic receptors regulate acetylcholine release in the central nervous system of the cockroach Periplaneta americana? J Exp Biol 1989; 142:447-451.
7. Le Corronc H, Lapied B, Hue B. M2-like presynaptic receptors modulate acetylcholine release in the cockroach (Periplaneta americana) central nervous system. J Insect Physiol 1991; 37:647-652.
8. Matsuda K, Buckingham SD, Kleier D et al. Neonicotinoids: Insecticides acting on insect nicotinic acetylcholine receptors. Trends Pharmacol Sci 2001; 22(11):573-80.
9. Honda H, Tomizawa M, Casida JE. Insect nicotinic acetylcholine receptors: Neonicotinoid binding site specificity is usually but not always conserved with varied substituents and species. J Agric Food Chem 2006; 54(9):3365-71.
10. Ihara M, Brown LA, Ishida C et al. Actions of imidacloprid, clothianidin and related neonicotinoids on nicotinic acetylcholine receptors of american cockroach neurons and their relationships with insecticidal potency. J Pestic Sci 2006; 31(1):35-40.
11. Breer H, Kleene R, Hinz G. Molecular forms and subunit structure of the acetylcholine rececptor in the central nervous of insects. J Neurosci 1985; 5:3386-3392.
12. Geffard M, Villemaringe J, Heinrich-Rock AM et al. Anti-acetylcholine antibodies and first immunocytochemical application in insect brain. Neurosci Lett 1985; 57:1-6.
13. Lutz EM, N.M. T. Immunohistochemical localization of choline acetyltransferase in the central nervous system of the locust. Brain Res 1987; 407:173-179.
14. Kreissl S, Bicker G. Histochemistry of acetylcholinesterase and immunocytochemistry of an acetylcholine receptor-like antigen in the brain of the honeybee. J Comp Neurol 1989; 286(1):71-84.
15. Scheidler A, Kaulen P, Bruning G et al. Quantitative autoradiographic localization of [^{125}I]alpha-bungarotoxin binding sites in the honeybee brain. Brain Res 1990; 534(1-2): 332-5.
16. Greenspan RJ. Mutations of choline acetyltransferase and associated neural defects in Drosophila melanogaster. J Comp Physiol 1980; 137:83-92.
17. Thany SH, Lenaers G, Raymond-Delpech V et al. Exploring the pharmacological properties of insect nicotinic acetylcholine receptors. Trends Pharmacol Sci 2007; 28(1):14-22.
18. Dudai Y, Amsterdam A. Nicotinic receptors in the brain of drosophila melanogaster demonstrated by autoradiography with [^{125}I]alpha-bungarotoxin. Brain Res 1977; 130(3):551-5.
19. Dudai Y. Properties of an α-bungarotoxin-binding cholinergic nicotinic receptor from Drosophila melanogaster. Biochim Biophys Acta 1978; 539:505-517.
20. Lummis SC, Sattelle DB. Binding of n-[propyonyl-3H] propionylated α-bungarotoxin and l-[benzilic-4,4-3H] quinuclidinyl benzilate to cns extracts of the cockroach Periplaneta americana. Comp Biochem Physiol C Comp Pharmacol 1985; 80:75-83.
21. Buckingham S, Lapied B, Corronc H et al. Imidacloprid actions on insect neuronal acetylcholine receptors. J Exp Biol 1997; 200(Pt 21):2685-92.
22. Hue B, Callec JJ. Electrophysiology and pharmacology of synaptic transmission in the central nervous system of the cockroach. In: Huber I, Masler EP, Rao BR, eds. Cockroaches as Models for Neurobiology: Applications in Biomedical Researches. Boca Raton: CRC Press, 1990:149-167.
23. Ngiam TL, GO ML. The structure of the acetylcholinesterase active center and the mode of acetylcholine/acetylcholinesterase interaction—a theoretical study. Asia Pacific J Pharmacol 1987; 2:33-41.
24. Toutant JP. Insect acetylcholinesterase: Catalytic properties, tissue distribution and molecular forms. Prog Neurobiol 1989; 32(5):423-46.
25. Hellenbrand K, Krupka RM. Kinetic studies on the mechanism of insect acetylcholinesterase. Biochem 1970; 9:4665-4672.
26. Toutant JP, Massoulié J, Bon S. Polymorphism of pseudocholinesterase in torpedo marmorata tissues: Comparative study of the catalytic and molecular properties of this enzyme with acetylcholinesterase. J Neurochem 1985; 44:580-592.
27. Gnagey AL, Forte M, Rosenberry TL. Isolation and characterization of acetylcholinesterase from drosophila. J Biol Chem 1987; 262:13290-13298.
28. Massoulié J, Bon S. The molecular forms of cholinesterase and acetylcholinesterase in vertebrates. Ann Rev Neurosci 1982; 5:57-106.
29. Arpagaus M, Fournier D, Toutant JP. Analysis of acetylcholinesterase molecular forms during the development of Drosophila melanogaster. Evidence for the existence of an amphiphilic monomer. Insect Biochem 1988; 18(6):539-549.
30. Fournier D, Cuany A, Bride JM et al. Molecular polymorphism of head acetylcholinesterase from adult houseflies (Musca domestica l.). J Neurochem 1987; 49:1455-1461.

31. Toutant JP, Arpagaus M, Fournier D. Native molecular forms of head acetylcholinesterase from adult Drosophila melanogaster: Quaternary structure and hydrophobic character. J Neurochem 1988; 50:209-218.
32. Pitman RM. Nervous system. In: Kerkut G, Gilbert LI, eds. Comprehensive Insect Physiology, Biochemistry and Pharmacology, Vol. 11. Oxford: Pergamon Press, 1985:5-54.
33. Melanson SW, Chang-Hyong Y, Pezzementi ML et al. Characterization of acetylcholinesterase activity from Drosophila melanogaster. Comp Biochem Physiol C Comp Pharmacol 1985; 81(1):87-96.
34. Sattelle DB, Ho YW, Crawford GD et al. Immunocytochemical staining of central neurones in Periplaneta americana using monoclonal antibodies to cholineacetyltransferase. Tissue Cell 1986; 18:51-61.
35. Pahud G, Salem N, van de Goor J et al. Study of subcellular localization of membrane-bound choline acetyltransferase in drosophila central nervous system and its association with membranes. Eur J Neurosci 1998; 10(5):1644-53.
36. Slemmon JR. Sequence analysis of a proteolyzed site in drosophila choline acetyltransferase. J Neurochem 1989; 52(6):1898-904.
37. Sugihara H, Andrisani V, Salvaterra PM. Drosophila choline acetyltransferase uses a non-aug initiation codon and full length rna is inefficiently translated. J Biol Chem 1990; 265(35):21714-9.
38. Gorczyca MG, Hall JC. Immunohistochemical localization of choline acetyltransferase during development and in chats mutants of drosophila melanogaster. J Neurosci 1987; 7(5):1361-9.
39. Yasuyama K, Salvaterra PM. Localization of choline acetyltransferase-expressing neurons in drosophila nervous system. Microsc Res Tech 1999; 45(2):65-79.
40. Leitinger G, Simmons PJ. Cytochemical evidence that acetylcholine is a neurotransmitter of neurons that make excitatory and inhibitory outputs in the locust ocellar visual system. J Comp Neurol 2000; 416:345-355.
41. Rind FC, Leitinger G. Immunocytochemical evidence that collision sensing neurons in the locust visual system contain acetylcholine. J Comp Neurol 2000; 423(3):389-401.
42. Python F, Stocker RF. Immunoreactivity against choline acetyltransferase, gamma-aminobutyric acid, histamine, octopamine and serotonin in the larval chemosensory system of Drosophila melanogaster. J Comp Neurol 2002; 453(2):157-67.
43. Clark J, Meisner S, Torkkeli PH. Immunocytochemical localization of choline acetyltransferase and muscarinic ach receptors in the antenna during development of the sphinx moth Manduca sexta. Cell Tissue Res 2005; 320(1):163-73.
44. Torkkeli PH, Widmer A, Meisner S. Expression of muscarinic acetylcholine receptors and choline acetyltransferase enzyme in cultured antennal sensory neurons and nonneural cells of the developing moth Manduca sexta. J Neurobiol 2005; 62(3):316-29.
45. Lutz EM, Lloyd SJ, Tyrer NM. Purification of choline acetyltransferase from the locust Schistocerca gregaria and production of serum antibodies to this enzyme. J Neurochem 1988; 50(1):82-9.
46. Fabian R, Seyfarth EA. Acetylcholine and histamine are transmitter candidates in identifiable mechanosensitive neurons of the spider Cupiennius salei: An immunocytochemical study. Cell Tissue Res 1997; 287(2):413-23.
47. Tucker ES, Tolbert LP. Reciprocal interactions between olfactory receptor axons and olfactory nerve glia cultured from the developing moth Manduca sexta. Dev Biol 2003; 260(1):9-30.
48. Breer H. Properties of putative nicotinic and muscarinic cholinergic receptors in the central nervous system of Locusta migratoria. Neurochem Int 1981; 3:43-52.
49. Breer H, Kleene R, Behnke D. Isolation of a putative nicotinic acetylcholine receptor from the central nervous system of Locusta migratoria. Neurosci Lett 1984; 46:323-328.
50. Sattelle DB, Breer H. Purification by affinity-chromatography of a nicotinic acetylcholine receptor from the cns of the cockroach Periplaneta americana. Comp Biochem Physiol C Comp Pharmacol 1985; 82(2):349-352.
51. Schmidt-Nielsen BK, Gepner JI, Teng NNH et al. Characterization of an alpha-bungarotoxin binding component from Drosophila melanogaster. J Neurochem 1977; 29:1013-1029.
52. Rudoff E. Acectylcholine receptors in the central nervous system of Drosophila melanogaster. Exp Cell Res 1978; 111:185-190.
53. Hildebrand JG, Hall LM, Osmond BC. Distribution of binding sites for ^{125}I-labeled alpha-bungarotoxin in normal and deafferented antennal lobes of Manduca sexta. Proc Natl Acad Sci USA 1979; 76(1):499-503.
54. Orr GL, Orr N, Hollingworth RM. Localization and pharmacological characterization of nicotinic-cholinergic binding sites in cockroach brain using alpha- and neuronal bungarotoxin. Insect Biochem 1990; 20:557-566.

55. Meyer MR, Reddy GR. Muscarinic and nicotinic cholinergic binding sites in the terminal abdominal ganglion of the cricket (Acheta domesticus). J Neurochem 1985; 45(4):1101-12.
56. Lind RJ, Clough MS, Earley FGP et al. Characterisation of multiple alpha-bungarotoxin binding sites in the aphid Myzus persicae (hemiptera: Aphididae). Insect Biochem Mol Biol 1999; 29:979-988.
57. Davies AR, Hardick DJ, Blagbrough IS et al. Characterisation of the binding of [^{3}H]methyllycaconitine: A new radioligand for labelling alpha 7-type neuronal nicotinic acetylcholine receptors. Neuropharmacology 1999; 38(5):679-90.
58. Lind RJ, Hardick DJ, Blagbrough IS et al. [^{3}H]-methyllycaconitine: A high affinity radioligand that labels invertebrate nicotinic acetylcholine receptors. Insect Biochem Mol Biol 2001; 31(6-7):533-42.
59. Lind RJ, Clough MS, Reynolds SE et al. [^{3}H]imidacloprid labels high- and low-affinity nicotinic acetylcholine receptor-like binding sites in the aphid Myzus persicae (hemiptera: Aphididae). Pest Biochem Physiol 1998; 62:3-14.
60. Grauso M, Reenan RA, Culetto E et al. Novel putative nicotinic acetylcholine receptor subunit genes, dalpha5, dalpha6 and dalpha7, in Drosophila melanogaster identify a new and highly conserved target of adenosine deaminase acting on RNA-mediated A-to-I pre-mRNA editing. Genetics 2002; 160(4):1519-33.
61. Jones AK, Marshall J, Blake AD et al. Sgbeta1, a novel locust (Schistocerca gregaria) non-alpha nicotinic acetylcholine receptor-like subunit with homology to the Drosophila melanogaster dbeta1 subunit. Invert Neurosci 2005; 5(3-4):147-155.
62. Jones AK, Grauso M, Sattelle DB. The nicotinic acetylcholine receptor gene family of the malaria mosquito, Anopheles gambiae. Genomics 2005; 85(2):176-87.
63. Jones AK, Raymond-Delpech V, Thany SH et al. The nicotinic acetylcholine receptor gene family of the honey bee, Apis mellifera. Genome Res 2006; 16(11):1422-30.
64. Jones AK, Sattelle DB. The cys-loop ligand-gated ion channel gene superfamily of the red flour beetle, Tribolium castaneum. BMC Genomics 2007; 8:327.
65. Chamaon K, Smalla KH, Thomas U et al. Nicotinic acetylcholine receptors of drosophila: Three subunits encoded by genomically linked genes can co-assemble into the same receptor complex. J Neurochem 2002; 80(1):149-57.
66. Chamaon K, Schulz R, Smalla KH et al. Neuronal nicotinic acetylcholine receptors of Drosophila melanogaster: The alpha-subunit dalpha3 and the beta-type subunit ard co-assemble within the same receptor complex. FEBS Lett 2000; 482(3):189-92.
67. Fayyazuddin A, Zaheer MA, Hiesinger PR et al. The nicotinic acetylcholine receptor dalpha7 is required for an escape behavior in drosophila. PLoS Biol 2006; 4(3):e63.
68. Hess N, Merz B, Gundelfinger ED. Acetylcholine receptors of the drosophila brain: A 900 bp promoter fragment contains the essential information for specific expression of the ard gene in vivo. FEBS Lett 1994; 346(2-3):135-40.
69. Gao JR, Deacutis JM, Scott JG. The nicotinic acetylcholine receptor subunit mdalpha6 from Musca domestica is diversified via posttranscriptional modification. Insect Mol Biol 2007; 16(3):325-34.
70. Hermsen B, Stetzer E, Thees R et al. Neuronal nicotinic receptors in the locust locusta migratoria. Cloning and expression. J Biol Chem 1998; 273(29):18394-404.
71. Thany SH, Lenaers G, Crozatier M et al. Identification and localization of the nicotinic acetylcholine receptor alpha3 mRNA in the brain of the honeybee, Apis mellifera. Insect Mol Biol 2003; 12(3):255-62.
72. Thany SH, Crozatier M, Raymond-Delpech V et al. Apisalpha2, apisalpha7-1 and apisalpha7-2: Three new neuronal nicotinic acetylcholine receptor alpha-subunits in the honeybee brain. Gene 2005; 344:125-32.
73. Jonas P, Baumann A, Merz B et al. Structure and developmental expression of the d alpha 2 gene encoding a novel nicotinic acetylcholine receptor protein of Drosophila melanogaster. FEBS Lett 1990; 269(1):264-8.
74. Wadsworth SC, Rosenthal LS, Kammermeyer KL et al. Expression of a Drosophila melanogaster acetylcholine receptor-related gene in the central nervous system. Mol Cell Biol 1988; 8(2):778-85.
75. Hermans-Borgmeyer I, Hoffmeister S, Sawruk E et al. Neuronal acetylcholine receptors in drosophila: Mature and immature transcripts of the ard gene in the devevloping central nervous system. Neuron 1989; 2:1147-1156.
76. Gundelfinger ED, Hess N. Nicotinic acetylcholine receptors of the central nervous system of drosophila. Biochim Biophys Acta 1992; 1137(3):299-308.
77. Sawruk E, Udri C, Betz H et al. Sbd, a novel structural subunit of the drosophila nicotinic acetylcholine receptor, shares its genomic localization with two alpha-subunits. FEBS Lett 1990; 273(1-2):177-81.

78. Fuhrman B, Partoush A, Aviram M. Acectylcholine esterase protects LDL against oxidation. Biochem Biophys Res Com 2004; 322:974-978.
79. Inestrosa NC, Alvarez A, Pérez CA et al. Acetylcholinesterase accelerates assembly of amyloid-beta-peptides into Alzheimer's fibrils: possible role of the peripheral site of the enzyme. Neuron 1996; 16(4):881-891.
80. Massoulié J, Millard CB. Cholinesterases and the basal lamina at vertebrate neuromuscular junctions. Curr Opin Pharmacol 2009; 9(3):316-325.

Chapter 2

The Evolution of Pentameric Ligand-Gated Ion Channels

Joseph A. Dent*

Abstract

Fast, ionotropic neurotransmission mediated by ligand-gated ion channels is essential for timely behavioral responses in multicellular organisms. Metazoa employ more ionotropic neurotransmitters in more types of synapses, inhibitory or excitatory, than is generally appreciated. It is becoming increasingly clear that the adaptability of a single neurotransmitter receptor superfamily, the pentameric ligand-gated ion channels (pLGICs), makes the diversity in ionotropic neurotransmission possible. Modification of a common pLGIC structure generates channels that are gated by ligands as different as protons, histamine or zinc and that pair common neurotransmitters with both cation and anion permeability. A phylogeny of the pLGIC gene family from representative metazoa suggests that pLGIC diversity is ancient and evolution of contemporary phyla was characterized by a surprising loss of pLGIC diversity. The pLGIC superfamily reveals aspects of early metazoan evolution, may help us identify novel neurotransmitters and can inform our exploration of structure/function relationships.

Introduction

With the cloning of the alpha subunit of the nicotinic acetylcholine receptor (nAChR), the nAChR became the exemplar of a functionally diverse superfamily of neurotransmitter receptors, the pentameric ligand-gated ion channels (pLGICs).[1] It also marked the beginning of the molecular analysis of neurotransmitter receptors and of our ability to understand channel structure, function and evolution in a new and enlightening dimension.[2,3] More than 25 years later, the pLGIC superfamily continues to expand with the discovery of new invertebrate and even bacterial pLGICs. Here I briefly summarize what is known about the underlying pLGIC structure and its many functional adaptations, I discuss some features of the pLGIC superfamily's evolution and finally I speculate on the opportunities that such a large and varied family of proteins presents for understanding metazoan evolution, neurochemistry and protein structure/function relationships.

Structure: pLGICs Share an Underlying Structure

One expects members of a protein superfamily to share structural motifs. The pLGIC superfamily combines a seemingly paradoxical lack of conservation at the level of sequence with a highly conserved underlying structure. The pLGIC channels are pentamers formed by five homologous subunits (Fig. 1). Channels can be homomeric or heteromeric with up to five different subunits

*Joseph A. Dent—Department of Biology, McGill University, 1205 Dr. Penfield Ave., Montreal QC H3A 1B1, Canada. Email: joseph.dent@mcgill.ca

Insect Nicotinic Acetylcholine Receptors, edited by Steeve Hervé Thany.

required to form a functional channel.[4] The amino acid sequence and predicted topology of the pLGIC subunits has traditionally defined their inclusion in a superfamily.[5-7] Even though the more divergent subunits show as little as 10-15% amino acid identity across a ~400-500 amino acid peptide, in multiple sequence alignments several residues are identical across almost the entire protein superfamily and scores of other residues show a high degree of conservation.[7,8] These conserved residues are scattered about the polypeptide, which facilitates sequence alignments and suggests that a characteristic structure underlies all pLGICs. That said, our criteria for what defines a pLGIC continue to be challenged as new pLGICs are characterized. For instance, a pair of what had appeared to be absolutely conserved cysteines that form a disulfide bond in the extracellular domain of pLGICs—and the reason metazoan pLGICs are referred to as "cys-loop" channels—turn out not to be present in bacterial pLGICs nor in at least one nematode pLGIC[7] (J. Dent. unpublished).

Conservation is not uniform throughout the subunit sequence. Based on topology and hydrophobicity, subunits can be divided into domains: a signal sequence, an amino-terminal extracellular ligand-binding domain, four transmembrane domains that form the transmembrane pore and, between the third and fourth transmembrane domains, a large cytoplasmic loop (Fig. 1A). The transmembrane domains are the most conserved, followed by the extracellular ligand-binding domain. The intracellular loops, even among closely related subunit types, typically show very little conservation.

In the past two decades we have been treated to increasingly detailed images that reveal the underlying pLGIC structure implied by the sequence conservation. These consist primarily of the electron diffraction images of quasicrystalline nAChRs from the electric ray *Torpedo californica*,[9-13] X-ray crystallographic data from the homopentameric acetylcholine binding protein (AChBP)—essentially a soluble pentameric extracellular domain without the transmembrane domains—from the snail *Lymnea stagnalis*,[14] crystallographic data of the extracellular domain of a mouse α1 nAChR subunit[15] and crystallographic data from the complete pentameric channels at 2.9-3.3 Å resolution from two bacterial pLGICs.[16-18] In these images the subunits traverse the membrane perpendicularly and form a ring that defines the transmembrane pore (Fig. 1B). When viewed from above, the channels appear much like a simple rose window or oculus. The extracellular domain is globular, extends ~30 Å above the lipid bilayer and consists primarily of beta sheets in the form of a beta sandwich. The extracellular domains sit atop the four transmembrane domains that are largely embedded in the membrane (Fig. 1C). The transmembrane domains of each channel subunit are alpha helices that span the lipid bilayer roughly perpendicular to the membrane surface. The M2 domains of each subunit line the pore with a stripe of polar amino acids on the M2 alpha helix facing the pore lumen. The M2 domains are circumscribed by a ring of alternating M1 and M3 domains, with the M4 domains deployed on the outside facing the lipid. The pore, starting at the lip of the extracellular domains, is funnel-shaped with the spout at the cytoplasmic end of the transmembrane domains. The short loop between the M2 and M3 domains on the extracellular side interacts with the disulfide ("cys") loop of the extracellular domains. In low resolution images of the nAChRs, the large intracellular loop, absent in bacterial pLGICs, forms a cytoplasmic vestibule through which ions must pass after escaping the transmembrane pore (Fig. 1C).[11]

The action of pLGICs begins when ligand-binding initiates a series of allosteric changes that propagate through the protein resulting in movement of the transmembranes domains and opening of the pore. Neurotransmitter ligands bind the extracellular domain, which is the determinant of ligand specificity.[19] We do not yet have a high-resolution structure of a pLGIC bound to its native ligand, but AChBP structures with bound agonists[20] confirm affinity labeling[21-26] and site-directed mutagenesis[27-31] studies showing that the ligand binding site sits at the interface between two subunits near the outer surface of the channel. A channel can have up to five ligand-binding sites depending on the how many adjacent subunit pairs bind ligand. The

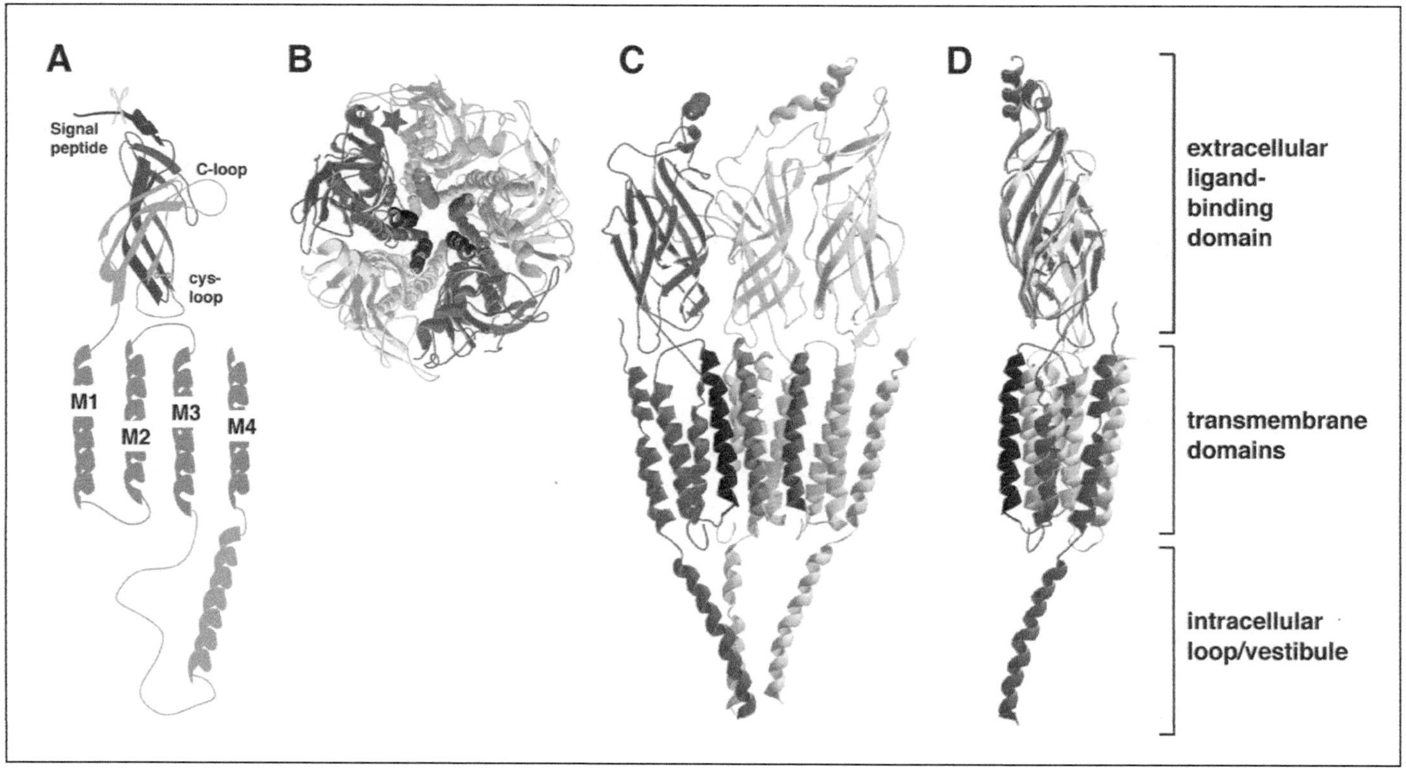

Figure 1. pLGIC structure. A) A cartoon showing the topology of a pLGIC subunit. B) A view from the extracellular side of the *Torpedo* nAChR looking down through the pore. Each subunit is rendered in a different shade. The M2 transmembrane domains that line the pore are darkened. The star indicates the location of a ligand binding site. C) A side view of B with two subunits removed to reveal the pore. Only part of the intracellular vestibule is resolved. D) One subunit from C (dark blue) overlain with the *Gloeobactor violaceus* pH-sensitive channel (GLIC, light lavender). Note the high degree of structural conservation despite low (~20%) sequence identity. A color version of this image is available at www.landesbioscience.com/curie.

ligand-binding pocket is a cage formed by aromatic side chains. The aromatic residues that form the pocket reside on four loops: the A, B and C loops on the (+) face—arbitrarily defined—of one subunit and the D loop on the (−) face of an adjacent subunit. Additional loops (E and F) on the (−) face also contribute to the binding site. In addition to delimiting the binding pocket, residues of the aromatic cage form cation-π interactions with the ligand, residues on different loops forming the cation-π interactions in different channels.[32-35] Apart from predictions based on homology models[27,36,37] and the demonstration that a phenylalanine to tyrosine change in the B-loop increases the sensitivity of glycine receptors to GABA,[38] little else is known about how pLGICs discriminate among ligands.

The transmembrane domains form the channel gate and the ion selective filter. The location of the gate and how the transmembrane domains move to open and close it remain controversial. Proposed gate locations on the M2 domain include: near the extracellular side[18] in the middle[9,39,40] or near the cytoplasmic side of the pore.[41] During gating the M2 domains may tilt or rotate.[9,17,18] In contrast, structural determinants of ion selectivity are well understood. pLGICs are typically selective for monovalent anions or cations. Rings of charged amino acids at the cytoplasmic side of the M2 domain, where it links to the M1 domain, determine the ion selectivity of the open pore. Changing the ion selectivity of a channel from anion to cation or vice versa is as simple as changing a few, or in some cases a single amino acid in the M2 domain.[42-45]

There has been significant progress tracing the allosteric steps between ligand-binding and gating of the pore. Results of mutagenesis experiments agree with predictions from crystal structures that the point of allosteric contact between the extracellular domain and the transmembrane domains occurs at the interface of the disulfide loop on the extracellular domain and the M2-M3 loop.[46-48] Using site-directed mutagenesis and single-channel kinetics, Auerbach and colleagues proposed a Brownian wave model of channel opening that involves sequential movement of four blocks of amino acids.[49,50] Thus, in a plausible model of channel gating, ligand-induced twisting of the beta sandwich nudges the M3-M4 loop tilting the M2 domains and opening the channel.

Even with only a handful of crystals, the conservation of the underlying pLGIC structure is striking. The representative extracellular domains from vertebrate, mollusk (AChBP) and bacterial subunits correspond almost perfectly at the level of secondary and tertiary structure despite the differences in function (pH response in bacteria[51] versus acetylcholine binding in metazoans) and scant primary sequence homology (20% amino acid identity in Fig. 1D). Presumably, all members of the pLGIC superfamily will adhere closely to the paradigm established by the known structures; differences in pLGIC properties will reflect modest adaptations of that paradigm.

Function: pLGICs Can Mediate Many Types of Ionotropic Neurotransmission

In metazoa, the biological function of the characterized pLGICs is to link neurotransmitter release by a presynaptic cell to ion permeability in a postsynaptic cell, thus converting a chemical signal to an electrical signal. However, the neurotransmitters and permeable ions can vary greatly. The ionotropic neurotransmitters in vertebrates are acetylcholine, GABA, glutamate, glycine, serotonin and adenosine triphosphate (ATP). Vertebrate pLGICs mediate neurotransmission by all but glutamate and ATP: glutamate is mediated by the structurally distinct tetrameric AMPA/NMDA-type receptors and ATP is mediated by trimeric P2X receptors. Vertebrate acetylcholine and serotonin receptors are cation-selective and thus excitatory. The GABA and glycine receptors are anion selective and typically inhibitory, although under certain conditions—early in development or after injury—the reversal potential of chloride, the primary permeable anion, can be high enough that GABA and glycine are excitatory.[52,53]

Compared to vertebrates, invertebrates employ a greatly expanded repertoire of neurotransmitter-ion combinations. In addition to inhibitory GABA receptors and excitatory nAChRs, there is evidence for: glutamate-gated chloride channels in insect muscle and in mollusk and crustacean neurons,[54-57] excitatory GABA currents in crustacean, mollusk and nematode muscle,[58-61] inhibitory histamine currents in the insect visual system,[62,63] inhibitory acetylcholine and dopamine currents as well as both excitatory and inhibitory histamine currents in mollusk neurons.[64-68] Although pharmacological evidence suggested that some of these currents are mediated by pLGICs,[69] proof that pLGICs are diverse enough to mediate all of these currents awaited cloning of the receptor subunits and examination of their primary sequence.

The extent of the corresponding pLGIC diversity is increasingly apparent, largely as a result of the sequencing of invertebrate genomes and the discovery of scores of predicted pLGIC subunits that are not obvious orthologs of the known vertebrate receptors. Characterization of these orphan pLGIC receptors led to the discovery of nematode acetylcholine-gated chloride channels,[70] insect cation-selective GABA receptors,[71] arthropod pH-gated chloride channels,[72,73] insect histamine gated chloride channels,[74,75] the mollusk acetylcholine-gated chloride channel[76] and a vertebrate Zn^{+}-gated cation channel.[77] Other channels found in directed searches for known currents include nematode and insect glutamate-gated chloride channels.[78,79] In the nematode *C. elegans*, new pLGICs have turned up in genetic screens for mutants affecting neurotransmission. These include the glutamate-gated chloride channels,[80,81] the GABA-gated cation channel,[82] the serotonin-gated chloride channel[83] and the pH-gated cation channel.[84]

Because of convergent evolution, a list of channel types based on ligand-ion combinations in fact under-represents the full extent of pLGIC diversity. Several channels share neurotransmitter-sensitivity and ion-selectivity but their subunits are clearly not orthologous. Examples include the glutamate-gated chloride channels from mollusks and ecdysozoa[85] and the acetylcholine-gated chloride channels from mollusks and nematodes[70,76] (Fig. 2). The mollusk acetylcholine-gated chloride channel appears to be a typical, if more divergent, nAChR and even shares with nAChRs a sensitivity to bungarotoxin.[76] However, its ion selectivity motif is that of an anion channel. In contrast, the sequence of the nematode acetylcholine-gated chloride channel subunit is more similar to other anion-selective (GABA-type) pLGIC subunits and it appears to have independently evolved the ability to bind acetylcholine. In fact, the ligand-binding site of nematode acetylcholine-gated chloride channels and nAChRs are not any more similar than any other pair of binding sites, an indication that acetylcholine-gated chloride channels evolved a unique structural solution to the problem of binding acetylcholine.

The emphasis here has been on pLGIC diversity yet there is significant overlap in the spectrum of pLGICs found in the various phyla. pLGIC subunit types that appear to be universal to bilateria include two types of nAChRs, the α1- and α7-types, as well as GABA-gated chloride channels composed of both α- and β-type subunits (Fig. 2). An obvious question is whether it is possible to infer the conservation of a particular circuit or behavior based on the conservation of pLGIC types. So far it appears not. α1-type nAChRs act at cholinergic neuromuscular junctions in vertebrates and nematodes whereas glutamate is the neuromuscular transmitter in insects.[86] α7-type nAChRs act in the vertebrate central nervous system and in macrophages[87] but can act at neuromuscular junctions in nematodes.[88] Thus, the context in which a given pLGIC type is used is highly adaptable.

Although the characterized metazoan pLGICs mediate neurotransmission, the discovery of pLGICs in prokaryotes hints at other possible roles. The GLIC (Gloebacter violaceus Ligand-gated Ion Channel) and ELIC (Erwinia chrysanthemi Ligand-gated Ion Channel) pLGICs are found in Gram-negative bacteria, which regulate pH and ion concentrations in the periplasmic space between the inner and outer membrane.[89] A reasonable but untested role for the bacterial pLGICs would be to maintain periplasmic ion homeostasis. If periplasmic channels are chemosensory, responding to cues in the environment, they may represent a first step on the path to the evolution of neurotransmitter receptors.[8,90] Some metazoan pLGICs may also play homeostatic or chemosensory roles that reflect the ancestral pLGIC function.[91,92]

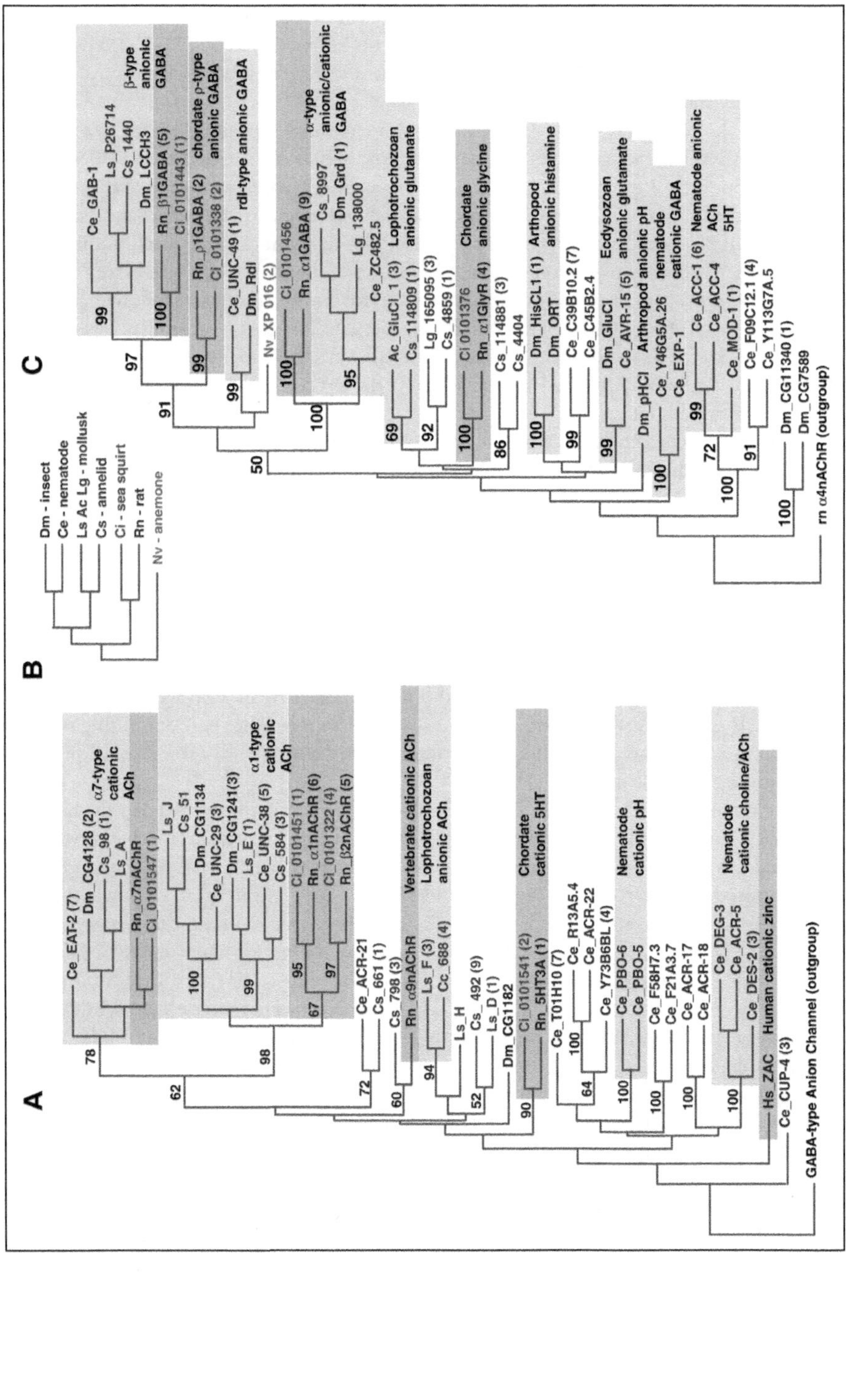

Figure 2. Please see legend on following page.

Figure 2, veiwed on previous page. Evolution of pLGIC subunits. A) A phylogenetic tree of representative nAChR-like channel subunits constructed using maximum likelihood. B) A phylogenetic tree showing evolutionary relationships of the species used in A and C. Ecdysozoa are represented by the clade that includes arthropods (insects) and nematodes, Lophotrochozoa by the clade that includes mollusks and annelids and chordates by the clade that includes rat and *Ciona*. C) A phylogenetic tree of representative GABA-like channel subunits. Subunits are from: *Aplysia californica* (Ac, mollusk), *Caenorhabditis elegans* (Ce, nematode), *Ciona intestinalis* (Ci, primitive chordate), *Capitella sp I* (Cs, annelid), *Drosophila melanogaster* (Dm, insect), *Homo sapiens* (Hs, vertebrate), *Lymnea stagnalis* (Ls, mollusk), *Lottia gigantea* (Lg, mollusk), *Nematostella vectensis* (Nv, anemone), *Rattus norvegicus* (Rn, vertebrate). Numbers on the branch indicate bootstrap values out of 100. Numbers in parentheses indicate subunits in that clade from that organism not shown in the tree. Subunits for which the ligand is known are shaded. Note that the list of subunits from *Ciona,* mollusks, annelids and *Nematostella* is incomplete.

Evolution: pLGIC Diversity Appears To Be Ancient

The diversity of pLGICs in metazoa begs the question how a protein superfamily of such scope evolved. Conspicuous and in need of explanation is the distinct spectrum of channels found in individual phyla.[8,85] Vertebrates have glycine receptors but ecdysozoa (nematodes and arthropods) and mollusks apparently lack them. Mollusks and ecdysozoa have glutamate-gated chloride channels but these are not present vertebrates. Even within the ecdysozoa, insects have histamine receptors but nematodes apparently do not, whereas nematodes have acetylcholine-gated chloride channels but insects do not. If these phylum- and superphylum-specific channels were not present in the common ancestor, when did they evolve? If they were in the common ancestor, when were they lost?

A phylogenetic analysis of the metazoan pLGIC superfamily suggests that almost all of the metazoan receptor types were likely present in the common ancestor of bilateria (protostomes and deuterostomes) and possibly in the common ancestor of the bilateria and cnidaria (jellyfish and anemones) as well[8,85,93] (Fig. 2). The evidence consists of the following observation and argument: all studied bilateria have pLGIC GABA receptors that are more closely related to each other than to any other channel type, i.e., they form a clade. Thus, the common ancestor of bilateria expressed a GABA receptor subunit and the divergence of the GABA receptor subunits reflects the radiation of the phyla that express them. Moreover, the GABA receptor subunit gene in the common ancestor of bilateria had already diverged from genes encoding other pLGIC subunit types. It follows that all of the other (nonGABA) pLGIC types found in any bilaterian were present in the common ancestor. The first cnidarian genome sequence, that of *Nematostella vectensis*, also appears to encode GABA receptors that diverged from the bilaterian GABA subunits after the various subunit types diverged,[85] which would push the origin of pLGIC diversity back to the common ancestor of the bilateria and cnidaria, an organism that existed more than 600 million years ago.[94,95]

The cation channels show a similar evolutionary pattern. There are several distinct types of nAChRs. The vertebrate α1- and β-type nAChR subunits typically combine to form obligate heteromeric channels.[96] In arthropods and nematodes there are two paralogous nAChR-like clades whose subunits also form heteromers.[4,97] Together, these four clades (two vertebrate and two invertebrate) form a larger clade (here referred to as the α1nAChR-like clade), meaning that the common ancestor of bilateria had at least one α1-type nAChR subunit.[8,98] A separate clade includes the vertebrate α7nAChR subunits as well as subunits from arthropods, nematodes, mollusks and annelids. Thus, the α7nAChR-like subunits also share an ancestral gene that predated the divergence of the bilateria. Beyond these shared clades are a number of phylum-specific clades: $5HT_3$ (cationic serotonin) receptors in vertebrates and several clades in invertebrates—so far primarily in the nematodes but also including the mollusk acetylcholine-gated chloride channel.[76] The nematode clades are little characterized but among them are the proton-gated cation channel[84] and a channel that responds most strongly to choline.[91] Whether the invertebrate cation channel clades are as diverse in ligand specificity as the anion channels remains to be seen, but in any case the tree topology supports a model in which the α1nAChR-like and α7nAChR-like subunits as well as the other cation channel subunit clades were present in the common ancestor of the bilateria.

The argument for a diverse pLGIC superfamily in the common ancestor of bilateria is subject to two caveats. The lesser caveat concerns the great phylogenetic/time span covered by trees that encompass all bilateria. Phylogenies measure distance in units of sequence change and unequal rates of change distort the distances between sequences. Nematodes evolve more rapidly than vertebrates[99] and therefore rate differences will distort the metazoan pLGIC phylogenies. But the argument for early divergence of subunit types depends on the topology of the tree, which does not seem to be affected. For instance, GABA receptors from nematodes and vertebrates form a clade, as do the nematode and vertebrate α7-type nAChRs, which they would not do if the apparently ancient origin of invertebrate-specific channels were an artifact of differences in the overall rate of evolution.[8] The second caveat is that the observed tree topologies could be explained by a recent (<500 myr) horizontal transfer of subunit genes. That explanation is difficult to support given the apparent rarity of horizontal gene transfer in metazoan.[100]

If the common ancestor of the bilateria had a more diverse complement of pLGICs than extant metazoa, gene loss must explain the difference. Primitive chordates such as *Ciona intestinalis* encode the same spectrum of pLGICs as vertebrates, consistent with the loss of invertebrate pLGICs before the divergence of the chordates. On the other hand, the modest overlap in nematode and arthropod pLGICs may indicate substantial gene loss since the divergence of the rapidly evolving ecdysozoa. Culling of pLGIC subunit genes from the various metazoan genomes has also not been monotonic. Certain subunit families have expanded substantially in some phyla and not others.[8,98] The vertebrate α1-type nAChRs appear to have expanded since the vertebrate-invertebrate split and even since the divergence of vertebrates from other chordates. The complement of GABA receptor subunits expanded substantially in vertebrates; where *C. elegans* and *Drosophila* have, respectively, two and three α- and β-type GABA receptor subunits, mammals have 19. Conversely, vertebrates encode a single α7nAChR subunit where *Drosophila* encodes three and *C. elegans* encodes nine. It seems the pLGIC superfamily is in constant flux. What drives the selective expansion and contraction of pLGIC subunit families is an intriguing and unexplored question.

What Good Is pLGIC Diversity?

The pLGIC superfamily is both a resource and an experimental system in which to explore questions of protein and nervous system evolution. There are three areas where I see the diversity of the pLGICs being exploited:

Pushing Back the Origin of Metazoan pLGICs

The sequencing of additional metazoan genomes raises the prospect of the being able to approximately reconstruct the genome of the common ancestor of the metazoa, including the component of the genome that encoded the pLGICs. The question is how far back we can trace the evolution of the existing channels before the obscuring effects of sequence divergence prevent any further deductions. Cnidaria apparently have both fast cholinergic and GABAergic neurotransmission and the next step will be to determine whether this neurotransmission is mediated by orthologs of the corresponding bilaterian pLGICs.[101] The genome sequence of *N. vectensis* will greatly facilitate that task. The next step would be to determine whether nAChRs and ionotropic GABA receptors are also present in a basal metazoan, a position proposed for the comb jellies.[102,103]

Identifying Novel Transmitters

With the advent of complete genome sequences it has been possible to catalog and characterize all of the pLGICs in an organism. The result has been the identification of pLGICs that respond to many neurotransmitters that were not previously thought to act on pLGICs. It has also resulted in the identification of pLGICs subunits that do not, at least in heterologous systems, respond to any known neurotransmitter (J. Dent unpublished observation). Many of these will be subunits that form obligate heteromeric channels and whose partner subunits have not been identified. However, where entire clades do not respond to known neurotransmitters,[8] we must consider the possibility that the in vivo ligand for these channels is not among the usual suspects. The recent discovery that protons mediate ionotropic neurotransmission via pLGICs demonstrates that characterization of orphan pLGIC receptors will lead us to new neurotransmitters.[84] One systematic approach to identify new neurotransmitters would be to use orphan pLGICs in a bioassay to screen candidate compounds.[104]

The Evolution of Ligand Specificity

The pLGIC superfamily represents at least half a billion years of evolutionary trial and error with the effect of identifying a large number of functional and physiologically useful neurotransmitter receptors from among the vastly larger number of useless pLGIC structures. The subunits that survive the evolutionary filter contain information about functional constraints on sequence and structure that can, in principle, be extracted using covariance techniques.[105,106] A rudimentary covariance approach was used to identify the residues that confer ion selectivity.[44] To identify subtler and more variable properties, such as ligand specificity, will require large sequence alignments from >100 functionally characterized channels, but we are rapidly approaching that number. Before us is the exciting prospect of combining correlative and crystallographic data to trace the path of amino acid changes that allowed the evolution of novel ligand specificities. Learning how to change ligand specificity is likely to be much more difficult than changing ion selectivity but also potentially more rewarding as it promises to enhance our understanding of the allosteric changes involved in gating. Ultimately, the test of our mastery of pLGIC structure, function and evolution will be to engineer channels with ligand specificities not found in nature for use as experimental probes, biosensors and medical therapies.

Acknowledgement

I would like to thank Ellis Cooper for helpful comments and discussion.

References

1. Noda M, Takahashi H, Tanabe T et al. Primary structure of alpha-subunit precursor of Torpedo californica acetylcholine receptor deduced from cDNA sequence. Nature 1982; 299(5886):793-797.
2. Noda M, Takahashi H, Tanabe T et al. Structural homology of Torpedo californica acetylcholine receptor subunits. Nature 1983; 302(5908):528-532.
3. Raftery MA, Hunkapiller MW, Strader CD et al. Acetylcholine receptor: complex of homologous subunits. Science 1980; 208(4451):1454-1456.
4. Boulin T, Gielen M, Richmond JE et al. Eight genes are required for functional reconstitution of the Caenorhabditis elegans levamisole-sensitive acetylcholine receptor. Proc Natl Acad Sci USA 2008; 105(47):18590-18595.
5. Schofield PR, Darlison MG, Fujita N et al. Sequence and functional expression of the GABA A receptor shows a ligand-gated receptor super-family. Nature 1987; 328(6127):221-227.
6. Grenningloh G, Rienitz A, Schmitt B et al. The strychnine-binding subunit of the glycine receptor shows homology with nicotinic acetylcholine receptors. Nature 1987; 328(6127):215-220.
7. Tasneem A, Lakshminaayan MI, Jakobsson E et al. Identification of the prokaryotic ligand-gated ion channels and their implications for the mechanisms and origins of animal Cys-loop ion channels. Genome Biol 2004; 6:R4.
8. Dent JA. Evidence for a diverse cys-loop ligand-gated ion channel superfamily in early bilateria. J Mol Evol 2006; in press.

9. Unwin N. Acetylcholine receptor channel imaged in the open state. Nature 1995; 373(6509):37-43.
10. Miyazawa A, Fujiyoshi Y, Unwin N. Structure and gating mechanism of the acetylcholine receptor pore. Nature 2003; 424(6943):949-955.
11. Unwin N. Refined structure of the nicotinic acetylcholine receptor at 4A resolution. J Mol Biol 2005; 346(4):967-989.
12. Toyoshima C, Unwin N. Three-dimensional structure of the acetylcholine receptor by cryoelectron microscopy and helical image reconstruction. J Cell Biol 1990; 111(6 Pt 1):2623-2635.
13. Toyoshima C, Unwin N. Ion channel of acetylcholine receptor reconstructed from images of postsynaptic membranes. Nature 1988; 336(6196):247-250.
14. Brejc K, van Dijk WJ, Klaassen RV et al. Crystal structure of an ACh-binding protein reveals the ligand-binding domain of nicotinic receptors. (see comment). Nature 2001; 411(6835):269-276.
15. Dellisanti CD, Yao Y, Stroud JC et al. Crystal structure of the extracellular domain of nAChR alpha1 bound to alpha-bungarotoxin at 1.94 A resolution. (see comment)(erratum appears in Nat Neurosci 2007; 10(9):1222). Nat Neurosci 2007; 10(8):953-962.
16. Hilf RJC, Dutzler R. X-ray structure of a prokaryotic pentameric ligand-gated ion channel. Nature 2008; 452(7185):375-379.
17. Bocquet N, Nury H, Baaden M et al. X-ray structure of a pentameric ligand-gated ion channel in an apparently open conformation. Nature 2009; 457(7225):111-114.
18. Hilf RJC, Dutzler R. Structure of a potentially open state of a proton-activated pentameric ligand-gated ion channel. Nature 2009; 457(7225):115-118.
19. Eisele JL, Bertrand S, Galzi JL et al. Chimaeric nicotinic-serotonergic receptor combines distinct ligand binding and channel specificities. (see comment). Nature 1993; 366(6454):479-483.
20. Celie PHN, van Rossum-Fikkert SE, van Dijk WJ et al. Nicotine and carbamylcholine binding to nicotinic acetylcholine receptors as studied in AChBP crystal structures. (see comment). Neuron 2004; 41(6):907-914.
21. Middleton RE, Cohen JB. Mapping of the acetylcholine binding site of the nicotinic acetylcholine receptor: (3H)nicotine as an agonist photoaffinity label. Biochemistry 1991; 30(28):6987-6997.
22. Galzi JL, Revah F, Black D et al. Identification of a novel amino acid alpha-tyrosine 93 within the cholinergic ligands-binding sites of the acetylcholine receptor by photoaffinity labeling. Additional evidence for a three-loop model of the cholinergic ligands-binding sites. J Biol Chem 1990; 265(18):10430-10437.
23. Froehner SC, Karlin A, Hall ZW. Affinity alkylation labels two subunits of the reduced acetylcholine receptor from mammalian muscle. Proc Natl Acad Sci USA 1977; 74(10):4685-4688.
24. Damle VN, McLaughlin M, Karlin A. Bromoacetylcholine as an affinity label of the acetylcholine receptor from Torpedo californica. Biochem Biophys Res Commun 1978; 84(4):845-851.
25. Weill CL, McNamee MG, Karlin A. Affinity-labeling of purified acetylcholine receptor from Torpedo californica. Biochem Biophys Res Commun 1974; 61(3):997-1003.
26. Smith GB, Olsen RW. Identification of a (3H)muscimol photoaffinity substrate in the bovine gamma-aminobutyric acidA receptor alpha subunit. J Biol Chem 1994; 269(32):20380-20387.
27. Grudzinska J, Schemm R, Haeger S et al. The beta subunit determines the ligand binding properties of synaptic glycine receptors. Neuron 2005; 45(5):727-739.
28. Galzi JL, Bertrand D, Devillers-Thiery A et al. Functional significance of aromatic amino acids from three peptide loops of the alpha 7 neuronal nicotinic receptor site investigated by site-directed mutagenesis. FEBS Letters 1991; 294(3):198-202.
29. Boileau AJ, Evers AR, Davis AF et al. Mapping the agonist binding site of the GABAA receptor: evidence for a beta-strand. J Neurosci 1999; 19(12):4847-4854.
30. Westh-Hansen SE, Witt MR, Dekermendjian K et al. Arginine residue 120 of the human GABAA receptor alpha 1, subunit is essential for GABA binding and chloride ion current gating. Neuroreport 1999; 10(11):2417-2421.
31. Wagner DA, Czajkowski C. Structure and dynamics of the GABA binding pocket: A narrowing cleft that constricts during activation. J Neurosci 2001; 21(1):67-74.
32. Padgett CL, Hanek AP, Lester HA et al. Unnatural amino acid mutagenesis of the GABA(A) receptor binding site residues reveals a novel cation-pi interaction between GABA and beta 2Tyr97. J Neurosci 2007; 27(4):886-892.
33. Lummis SCR, L Beene D, Harrison NJ et al. A cation-pi binding interaction with a tyrosine in the binding site of the GABAC receptor. Chem Biol 2005; 12(9):993-997.
34. Beene DL, Brandt GS, Zhong W et al. Cation-pi interactions in ligand recognition by serotonergic (5-HT3A) and nicotinic acetylcholine receptors: the anomalous binding properties of nicotine. Biochemistry 2002; 41(32):10262-10269.

35. Pless SA, Millen KS, Hanek AP et al. A cation-pi interaction in the binding site of the glycine receptor is mediated by a phenylalanine residue. J Neurosci 2008; 28(43):10937-10942.
36. Le Novere N, Grutter T, Changeux J-P. Models of the extracellular domain of the nicotinic receptors and of agonist- and Ca^{2+}-binding sites. Proc Natl Acad Sci USA 2002; 99(5):3210-3215.
37. Harrison NJ, Lummis SCR. Molecular modeling of the GABA(C) receptor ligand-binding domain. J Mol Model 2006; 12(3):317-324.
38. Schmieden V, Kuhse J, Betz H. Mutation of glycine receptor subunit creates beta-alanine receptor responsive to GABA. Science 1993; 262(5131):256-258.
39. Bali M, Akabas MH. The location of a closed channel gate in the GABAA receptor channel. J Gen Physiol 2007; 129(2):145-159.
40. Panicker S, Cruz H, Arrabit C et al. Evidence for a centrally located gate in the pore of a serotonin-gated ion channel. J Neurosci 2002; 22(5):1629-1639.
41. Wilson GG, Karlin A. The location of the gate in the acetylcholine receptor channel. Neuron 1998; 20(6):1269-1281.
42. Jensen ML, Timmermann DB, Johansen TH et al. The beta subunit determines the ion selectivity of the GABAA receptor. J Biol Chem 2002; 277(44):41438-41447.
43. Gunthorpe MJ, Lummis SC. Conversion of the ion selectivity of the 5-HT(3a) receptor from cationic to anionic reveals a conserved feature of the ligand-gated ion channel superfamily. J Biol Chem 2001; 276(24):10977-10983.
44. Galzi JL, Devillers-Thiery A, Hussy N et al. Mutations in the channel domain of a neuronal nicotinic receptor convert ion selectivity from cationic to anionic. Nature 1992; 359(6395):500-505.
45. Corringer PJ, Bertrand S, Galzi JL et al. Mutational analysis of the charge selectivity filter of the alpha7 nicotinic acetylcholine receptor. Neuron 1999; 22(4):831-843.
46. Lummis SCR, Beene DL, Lee LW et al. Cis-trans isomerization at a proline opens the pore of a neurotransmitter-gated ion channel. (see comment). Nature 2005; 438(7065):248-252.
47. Kash TL, Jenkins A, Kelley JC et al. Coupling of agonist binding to channel gating in the GABA(A) receptor. Nature 2003; 421(6920):272-275.
48. Grosman C, Salamone FN, Sine SM et al. The extracellular linker of muscle acetylcholine receptor channels is a gating control element. J Gen Physiol 2000; 116(3):327-340.
49. Auerbach A. Gating of acetylcholine receptor channels: brownian motion across a broad transition state. Proc Natl Acad Sci USA 2005; 102(5):1408-1412.
50. Purohit P, Mitra A, Auerbach A. A stepwise mechanism for acetylcholine receptor channel gating. Nature 2007; 446(7138):930-933.
51. Bocquet N, Prado de Carvalho L, Cartaud J et al. A prokaryotic proton-gated ion channel from the nicotinic acetylcholine receptor family. Nature 2007; 445(7123):116-119.
52. De Koninck Y. Altered chloride homeostasis in neurological disorders: a new target. Curr Opin Pharmacol 2007; 7(1):93-99.
53. Ben-Ari Y, Gaiarsa J-L, Tyzio R et al. GABA: a pioneer transmitter that excites immature neurons and generates primitive oscillations. Physiol Rev 2007; 87(4):1215-1284.
54. Cull-Candy SG. Two types of extrajunctional L-glutamate receptors in locust muscle fibres. Journal of Physiology 1976; 255(2):449-464.
55. Kehoe J, Vulfius C. Independence of and interactions between GABA-, glutamate- and acetylcholine-activated Cl conductances in Aplysia neurons. J Neurosci 2000; 20(23):8585-8596.
56. Marder E, Paupardin-Tritsch D. The pharmacological properties of some crustacean neuronal acetylcholine, gamma-aminobutyric acid and L-glutamate responses. J Physiol 1978; 280:213-236.
57. Usherwood PN, Cull-Candy SG. Distribution of glutamate sensitivity on insect muscle fibres. Neuropharmacology 1974; 13(6):455-461.
58. Norekian TP. GABAergic excitatory synapses and electrical coupling sustain prolonged discharges in the prey capture neural network of Clione limacina. J Neurosci 1999; 19(5):1863-1875.
59. McIntire SL, Jorgensen E, Kaplan J et al. The GABAergic nervous system of Caenorhabditis elegans. (see comment). Nature 1993; 364(6435):337-341.
60. Yarowsky PJ, Carpenter DO. Receptors for gamma-aminobutyric acid (GABA) on Aplysia neurons. Brain Res 1978; 144(1):75-94.
61. Swensen AM, Golowasch J, Christie AE et al. GABA and responses to GABA in the stomatogastric ganglion of the crab Cancer borealis. J Exp Biol 2000; 203(Pt 14):2075-2092.
62. Hardie RC. A histamine-activated chloride channel involved in neurotransmission at a photoreceptor synapse. Nature 1989; 339(6227):704-706.

63. Gengs C, Leung H-T, Skingsley DR et al. The target of Drosophila photoreceptors synaptic transmission is a histamine-gated chloride channel encoded by ort (hclA). J Biol Chem 2002; 277:42113-42120.
64. Kehoe J. Ionic mechanisms of a two-component cholinergic inhibition in Aplysia neurones. J Physiol 1972; 225(1):85-114.
65. McCaman RE, Weinreich D. Histaminergic synaptic transmission in the cerebral ganglion of Aplysia. J Neurophysiol 1985; 53(4):1016-1037.
66. Magoski NS, Bulloch AG. Dopamine activates two different receptors to produce variability in sign at an identified synapse. J Neurophysiol 1999; 81(3):1330-1340.
67. Green KA, Harris SJ, Cottrell GA. Dopamine directly activates a ligand-gated channel in snail neurones. Pflugers Archiv—Eur J Physiol 1996; 431(4):639-644.
68. Green KA, Lambert JJ, Cottrell GA. Ligand-gated ion channels opened by 5-HT in molluscan neurones. Brit J Pharmacol 1996; 119(3):602-608.
69. Kehoe J, McIntosh JM. Two distinct nicotinic receptors, one pharmacologically similar to the vertebrate alpha7-containing receptor, mediate Cl currents in aplysia neurons. J Neurosci 1998; 18(20):8198-8213.
70. Putrenko I, Zakikhani M, Dent JA. A family of acetylcholine-gated chloride channel subunits in Caenorhabditis elegans. J Biol Chem 2005; 280(8):6392-6398.
71. Gisselmann G, Plonka J, Pusch H et al. Drosophila melanogaster GRD and LCCH3 subunits form heteromultimeric GABA-gated cation channels. Br J Pharamcol 2004; 142(3):409-413.
72. Schnizler K, Saeger B, Pfeffer C et al. A novel chloride channel in Drosophila melanogaster is inhibited by protons. J Biol Chem 2005; 280(16):16254-16262.
73. Mounsey KE, Dent JA, Holt DC et al. Molecular characterisation of a pH-gated Chloride Channel from Sarcoptes scabiei. Invert Neurosci 2007; 7:149-156.
74. Gisselmann G, Pusch H, Hovemann BT et al. Two cDNAs coding for histamine-gated ion channels in D. melanogaster. Nat Neurosci 2002; 5(1):11-12.
75. Zheng Y, Hirschberg B, Yuan J et al. Identification of two novel Drosophila melanogaster histamine-gated chloride channel subunits expressed in the eye. J Biol Chem 2002; 277(3):2000-2005.
76. van Nierop P, Keramidas A, Bertrand S et al. Identification of molluscan nicotinic acetylcholine receptor (nAChR) subunits involved in formation of cation- and anion-selective nAChRs. J Neurosci 2005; 25(46):10617-10626.
77. Davies PA, Wang W, Hales TG et al. A novel class of ligand-gated ion channel is activated by Zn2+. J Biol Chem 2003; 278(2):712-717.
78. Cully DF, Paress PS, Liu KK et al. Identification of a Drosophila melanogaster glutamate-gated chloride channel sensitive to the antiparasitic agent avermectin. J Biol Chem 1996; 271(33):20187-20191.
79. Cully DF, Vassilatis DK, Liu KK et al. Cloning of an avermectin-sensitive glutamate-gated chloride channel from Caenorhabditis elegans. Nature 1994; 371(6499):707-711.
80. Dent JA, Davis MW, Avery L. avr-15 encodes a chloride channel subunit that mediates inhibitory glutamatergic neurotransmission and ivermectin sensitivity in worms. EMBO J 1997; 16(19):5867-5879.
81. Dent JA, Smith M, Vassilatis DK et al. The genetics of ivermectin resistance in Caenorhabditis elegans. Proc Natl Acad Sci USA 2000; 97(6):2674-2679.
82. Beg AA, Jorgensen EM. EXP-1 is an excitatory GABA-gated cation channel. (see comment). Nat Neurosci 2003; 6(11):1145-1152.
83. Ranganathan R, Cannon SC, Horvitz HR. MOD-1 is a serotonin-gated chloride channel that modulates locomotory behaviour in C. elegans. Nature 2000; 408(6811):470-475.
84. Beg AA, Ernstrom GG, Nix P et al. Protons act as a transmitter for muscle contraction in C. elegans. (see comment). Cell 2008; 132(1):149-160.
85. Kehoe J, Buldakova S, Acher F et al. Two glutamate-gated chloride channels from Aplysia californica cloned, expressed and subjected to phylogenetic and sequence analysis with vertebrate and recently-cloned invertebrate homologs 2009; Submitted.
86. Usherwood PN, Machili P, Leaf G. L-Glutamate at insect excitatory nerve-muscle synapses. Nature 1968; 219(5159):1169-1172.
87. Wang H, Yu M, Ochani M et al. Nicotinic acetylcholine receptor alpha7 subunit is an essential regulator of inflammation. (see comment). Nature 2003; 421(6921):384-388.
88. McKay JP, Raizen DM, Gottschalk A et al. eat-2 and eat-18 are required for nicotinic neurotransmission in the Caenorhabditis elegans pharynx. Genetics 2004; 166(1):161-169.
89. Stock JB, Rauch B, Roseman S. Periplasmic space in Salmonella typhimurium and Escherichia coli. J Biol Chem 1977; 252(21):7850-7861.

90. Cockroft VB, Osguthorpe EA, Barnard EA et al. Ligand-gated ion channels. Mol Neurobiol 1992; 4:129-169.
91. Yassin L, Boaz G, Kahan T et al. Characterization of the DEG-3/DES-2 receptor: a nicotinic acetylcholine receptor that mutates to cause neuronal degeneration. Mol Cell Neurosci 2001; 17:589-599.
92. Patton A, Knuth S, Schaheen B et al. Endocytosis function of a ligand-gated ion channel homolog in Caenorhabditis elegans. Curr Biol 2005; 15(11):1045-1050.
93. Xue H. Identification of major phylogenetic branches of inhibitory ligand-gated channel receptors. J Mol Evol 1998; 47:323-333.
94. Wray GA, Levinton JS, Shapiro LH. Molecular evidence for deep precambrian divergences among metazoan phyla. Science 1996; 274(5287):568-573.
95. Peterson KJ, Lyons JB, Nowak KS et al. Estimating metazoan divergence times with a molecular clock. Proc Natl Acad Sci USA 2004; 101(17):6536-6541.
96. Hille B. Ionic Channels of Excitable Membranes. 3rd ed. Sunderland, Mass.: Sinauer Associates, Inc.; 1992.
97. Richmond JE, Jorgensen EM. One GABA and two acetylcholine receptors function at the C. elegans neuromuscular junction. Nat Neurosci 1999; 2(9):791-797.
98. Ortells MO, Lunt GG. Evolutionary history of the ligand-gated ion-channel superfamily of receptors. Trends Neurosci 1995; 18(3):121-127.
99. Coghlan A, Wolfe KH. Fourfold faster rate of genome rearrangement in nematodes than in Drosophila. Genome Res 2002; 16:857-867.
100. Keeling PJ, Palmer JD. Horizontal gene transfer in eukaryotic evolution. Nat Rev Genet 2008; 9(8):605-618.
101. Kass-Simon G, Pierobon P. Cnidarian chemical neurotransmission, an updated overview. Comp Biochem Physiol A Mol Integr Physiol146(1):9-25.
102. Dunn CW, Hejnol A, Matus DQ et al. Broad phylogenomic sampling improves resolution of the animal tree of life. Nature 2008; 452(7188):745-749.
103. Chen J-Y, Schopf JW, Bottjer DJ et al. Raman spectra of a Lower Cambrian ctenophore embryo from southwestern Shaanxi, China. Proc Natl Acad Sci USA 2007; 104(15):6289-6292.
104. Martinez-Torres A, Miledi R. Expression of Caenorhabditis elegans neurotransmitter receptors and ion channels in Xenopus oocytes. Proc Natl Acad Sci USA 2006; 103(13):5120-5124.
105. Lockless SW, Ranganathan R. Evolutionarily conserved pathways of energetic connectivity in protein families. Science 1999; 286(5438):295-299.
106. Chen Y, Reilly K, Chang Y. Evolutionarily conserved allosteric network in the Cys loop family of ligand-gated ion channels revealed by statistical covariance analyses. J Biol Chem 2006; 281(26):18184-18192.

Chapter 3

Diversity of Insect Nicotinic Acetylcholine Receptor Subunits

Andrew K. Jones* and David B. Sattelle

Abstract

Nicotinic acetylcholine receptors (nAChRs) are ligand-gated ion channels that mediate fast synaptic transmission in the insect nervous system and are targets of a major group of insecticides, the neonicotinoids. They consist of five subunits arranged around a central ion channel. Since the subunit composition determines the functional and pharmacological properties of the receptor the presence of nAChR families comprising several subunit-encoding genes provides a molecular basis for broad functional diversity. Analyses of genome sequences have shown that nAChR gene families remain compact in diverse insect species, when compared to their nematode and vertebrate counterparts. Thus, the fruit fly (*Drosophila melanogaster*), malaria mosquito (*Anopheles gambiae*), honey bee (*Apis mellifera*), silk worm (*Bombyx mori*) and the red flour beetle (*Tribolium castaneum*) possess 10-12 nAChR genes while human and the nematode *Caenorhabditis elegans* have 16 and 29 respectively. Although insect nAChR gene families are amongst the smallest known, receptor diversity can be considerably increased by the posttranscriptional processes alternative splicing and mRNA A-to-I editing which can potentially generate protein products which far outnumber the nAChR genes. These two processes can also generate species-specific subunit isoforms. In addition, each insect possesses at least one highly divergent nAChR subunit which may perform species-specific functions. Species-specific subunit diversification may offer promising targets for future rational design of insecticides that target specific pest insects while sparing beneficial species.

Introduction

Since the groundbreaking sequencing of the first insect genome, that of the fruit fly *Drosophila melanogaster*, several other insect genomes have been sequenced allowing for detailed comparisons of gene families. In this chapter we explore the diversity of nicotinic acetylcholine receptor (nAChR) gene families in various insect species such as the fruit fly genetic model organism (*Drosophila melanogaster*), the malarial disease vector (*Anopheles gambiae*), the agriculturally beneficial honey bee (*Apis mellifera*), the commercially important silk worm (*Bombyx mori*) and the red flour beetle (*Tribolium castaneum*) which is a pest species of stored food. nAChRs are part of a ligand-gated ion channel superfamily found in species as diverse as bacteria and human and their best known role is molecular signalling in nervous systems and neuromuscular junctions as well as in nonneuronal cells. The central nervous system of insects is rich in nAChRs, more so than any other organism apart from the electroplax tissue of the electric fish. Insect

*Corresponding Author: Andrew K. Jones—MRC Functional Genomics Unit, Department of Physiology, Anatomy and Genetics, University of Oxford, Oxford, OX1 3QX, UK. Email: andrew.jones@dpag.ox.ac.uk

Insect Nicotinic Acetylcholine Receptors, edited by Steeve Hervé Thany.

nAChRs are therefore of interest for the study of nervous system signalling molecules and as targets for several classes of important insecticides.

Nicotinic Acetylcholine Receptors (nAChRs)—Structure and Function

Nicotinic acetylcholine receptors (nAChRs) are prototypical members of the Cys-loop ligand-gated ion channel (cysLGIC) superfamily[1,2] which also includes ionotropic receptors for GABA, glycine and serotonin (5-hydoxytryptamine). nAChRs mediate the fast actions of acetylcholine (ACh) in the nervous system and at neuromuscular junctions and consist of five homologous subunits arranged around a central ion channel (Fig. 1). They act as molecular switches which change conformation upon binding to an agonist such as ACh to allow a net influx of ions into the cell.[3] Each subunit has four transmembrane domains (TM1-4) and possesses an N-terminal extracellular domain containing the characteristic Cys-loop motif consisting of two disulfide bond-forming cysteines separated by 13 amino acid residues. The Cys-loop plays a role in nAChR assembly[4] as well the kinetics of ion channel gating.[5] The ACh-binding site is located at the interface of two adjacent subunits and is formed by six distinct regions (loops A-F)[6] in the N-terminal extracellular domain with loops A, B and C being contributed by an α subunit and loops D, E and F by either an α or non-α subunit. Subunits possessing two adjacent cysteine residues in loop C which are important for ACh binding[7] are defined as α subunits while subunits lacking these vicinal cysteines are classified as non-α (β, δ, ε or γ).

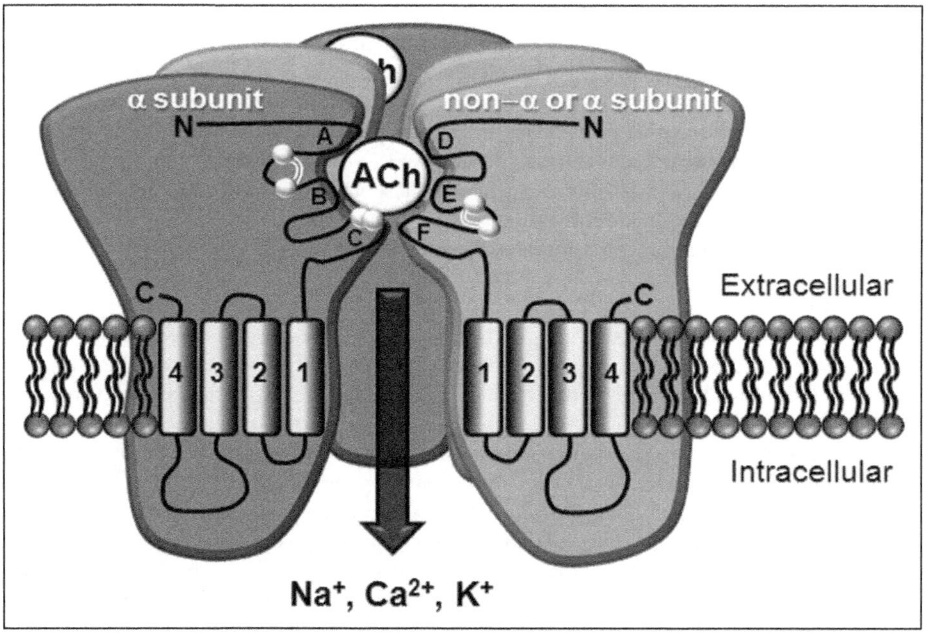

Figure 1. Structure of the nicotinic acetylcholine receptor. Schematic representation of a heteromeric receptor consisting of two α (dark grey) and three non-α subunits (light grey). The polypeptide layout of two subunits are shown highlighting the Cys-loop (two white circles connected by a white double line), the two vicinal cysteines in loop C defining α subunits and four transmembrane domains (TM1-4) with a large intracellular loop between TM3 and TM4. The six binding loops (A-F) that contribute to ligand binding are shown and two acetylcholine (ACh) molecules are bound to this particular nAChR. The five subunits that make up the receptor are arranged around a central cation-permeable channel.

nAChRs can exist as homomers of α subunits or as heteromers of either two kinds of α subunit, or, more commonly, of various combinations of α and non-α subunits.[8] The subunit composition determines the functional and pharmacological properties of the nAChR, thus receptor diversity is generated by multiple-subunit-encoding genes in a given organism.

Although no crystal structure of a full nAChR is currently available, the structure of the marine ray *Torpedo marmorata* electric organ nAChR has been resolved at 4.0 Å providing valuable insights into the three-dimensional structure of a cysLGIC.[9] More information has been added by further crystal structures such as that obtained at 1.94 Å for the N-terminal extracellular domain of the mouse α1 muscle subunit.[10] The α1 subunit is bound to α-bungarotoxin, a snake toxin which was used in the first purification of a nAChR[11] and the crystal structure has provided important insights into protein-protein and protein-sugar interactions of the subunit-toxin complex.[10] In addition, the crystal structure of an ACh binding protein (AChBP) from the pond snail *Lymnaea stagnalis* that shares homology with the extracellular N-terminal region of the nicotinic acetylcholine receptor has been determined at 2.7 Å.[12] Also, the X-ray structure of a bacterial LGIC with a layout of five subunits similar to cysLGICs has been determined at 3.3 Å[13] accelerating the exciting prospect of a crystal structure for a complete eukaryotic nAChR. These structures have considerably enhanced our understanding of receptor function by permitting the construction of three-dimensional homology models of cysLGICs and the computational simulation of receptor dynamics as well as agonist docking in the ligand-binding site.

Nicotinic Acetylcholine Receptors—Roles in Human Disease and as Drug Targets

The human nAChR family consists of 16 genes encoding 10 α and 6 non-α subunits.[8] There are separate families of muscle and neuronal nAChRs and the striking differences in pharmacological properties of nicotinic receptor subtypes found in different cells and tissues are mainly attributed to differences in their subunit composition. The importance of nAChRs is highlighted by their involvement in genetic and autoimmune disorders.[14] For instance, mutations in neuronal nAChR subunits α4 and β2 are associated with autosomal dominant nocturnal frontal lobe epilepsy[15] while mutations in muscle nAChR subunits (α, β, δ and ε) underlie muscle weakening congenital myasthenia syndromes.[16] Examples of autoimmune diseases include myasthenia gravis where auto antibodies target muscle nAChRs[17] while auto antibodies to α7 nAChRs, which function in the central nervous system, result in Rasmussen's encephalitis.[18] nAChRs play important roles in brain function and are the focus of research investigating them as targets for drugs designed to treat nicotine addiction, Alzheimer's disease, Parkinson's disease and Schizophrenia.[19]

Insect Nicotinic Acetylcholine Receptors—Targets for Pest Control

Invertebrate nAChRs are of interest as they are effective targets for pest control. Nematode (worm) parasites infect a billion people and also cause many serious diseases in livestock as well as crop damage. Levamisole, pyrantel and morantel are anthelmintics (drugs used to control worm parasites) which target nAChRs functioning in body wall muscles of nematodes[20] and recently a novel class of drugs (the amino-acetonitrile derivatives) has been developed which target a different nAChR subtype of nematodes.[21]

With ACh being an abundant neurotransmitter in the nervous systems of many insect species, including the fruit fly, *Drosophila melanogaster*,[22] nAChRs are targeted by chemicals used for insect control, such as neonicotinoids[23] which have been the fastest-growing class of insecticides in modern crop protection. Imidacloprid [1-(6-chloro-3-pyridylmethyl)-2-nitroimino-imidazolidine] and other neonicotinoids now have worldwide annual sales of around $US 1.56 billion, representing nearly 17% of the global insecticide market.[24] It has been demonstrated that imidacloprid binds with high affinity to membrane preparations from diverse insects.[25-27] Radioligand binding and electrophysiological studies have demonstrated that imidacloprid shows agonist actions on nAChRs in cockroach neurons and displaces ^{125}I α-bungarotoxin binding from central nervous system membranes.[28] Calcium imaging has demonstrated that imidacloprid acts on nAChRs in *D. melanogaster* cholinergic neurons[29] and whole-cell patch-clamp studies on the same neurons show imidacloprid to be a partial agonist.[30] The effectiveness of neonicotinoids as safe insecticides has been attributed, at least in part, to the selectivity for insect nicotinic receptors over mammalian nAChRs[31] and indeed the binding affinity of neonicotinoids to nAChRs correlates well with insecticidal efficacy.[32]

The nAChR Gene Family in a Genetic Model Organism, the Fruitfly *Drosophila melanogaster*

The First Complete Insect nAChR Gene Family To Be Described

The first sequences of nAChR subunits, those of the electric rays *Torpedo californica* and *Torpedo marmorata*, were published in the early 1980s.[33-37] Subsequently, it has been shown that nAChR subunits from a variety of organisms are highly homologous sharing considerable amino acid identity.[6] This enabled development of Torpedo DNA probes to isolate *D. melanogaster* cDNA clones in a hybridisation screen which led to the determination of the first insect nAChR subunit sequence.[38] As summarised in Table 1, nine further *D. melanogaster* nAChR subunits were identified over the

Table 1. Summary of all 10 D. melanogaster nAChR subunits which are listed in the order their sequences were published. The techniques used to determine their sequences are included

Subunit Name	Technique Used to Identify Subunit	Year Published
Dβ1 or ARD (acetylcholine receptor Drosophila)	Hybridisation screen using Torpedo α and γ nAChR subunit probes	1986[38]
Dα1 or ALS (alpha-like subunit)	Hybridisation screen using chicken α2 nAChR subunit probe	1988[39]
Dα2 or SAD (second alpha-like subunit Drosophila)	Hybridisation screen using Dα1 nAChR subunit probe[41] or conserved 10 amino acid region preceding TM4[44]	1990[41,44]
Dβ2 or SBD (second beta-like subunit Drosophila)	Hybridisation screen using genomic clone probe isolated in ref.[44]	1990[45]
Dα3	Hybridisation screen using Dα1 nAChR subunit probe and conserved 10 amino acid region preceding TM4	1998[46]
Dα4	PCR using primers based on an EST clone	2000[42]
Dβ3	Sequencing of an EST clone based on a gene predicted in the *D. melanogaster* genome	2002[43]
Dα5	BLAST analysis[124] against *D. melanogaster* genome sequence	2002[40]
Dα6	BLAST analysis[124] against *D. melanogaster* genome sequence	2002[40]
Dα7	BLAST analysis[124] against *D. melanogaster* genome sequence	2002[40]

next two decades.[38-46] During this time the *D. melanogaster* genome was sequenced[47] which greatly facilitated the identification of nAChR subunits, changing the strategy of isolating subunits from hybridisation screening to genome sequence analysis (Table 1). With the genome sequence available, it was shown that the complete fruit fly nAChR gene family consists of 10 subunits, seven of which are α (Dα1-Dα7) and three are non-α (Dβ1-Dβ3).[48] Considering that humans possess 16 subunits[8] and the nematode worm *Caenorhabditis elegans* has at least 29,[49] the nAChR gene family of *D. melanogaster* is rather compact. However, as described in section 4, alternative splicing and RNA editing considerably increases the number of insect nAChR gene products.

Distribution and Assembly of Drosophila AChRs

The localisation of many gene products throughout the Drosophila body can be determined by immunohistochemistry or in situ hybridisation. These techniques have been used to show that several Drosophila nAChR subunits (Dα1, Dα2, Dα3, Dα4, Dα7,[50] Dβ1 and Dβ2) have overlapping distributions in various regions of the nervous system (for review see ref. 96 and refs. therein). A lot is known about the subunit composition of vertebrate nAChRs.[8] Unfortunately, this is not the case for Drosophila, largely due to the fact that expression of functional receptors in heterologous systems has so far been unsuccessful. Several studies, however, have provided clues regarding the assembly and functions of certain subunits (for review see ref. 96 and refs. therein). Thus, based on immunoprecipitation experiments, overlapping expression patterns and pharmacological properties of hybrid receptors consisting of Drosophila α subunits and vertebrate non-α subunits, Chamaon et al proposed three possible receptor complexes.[51] One contains at least Dβ1 and Dβ2, another includes Dβ1 and Dα3 whilst in the third at least Dα1, Dα2 and Dβ2 are present. As noted by the authors, the genes encoding Dα1, Dα2 and Dβ2 form a directly linked cluster in the *Drosophila* genome which may facilitate coordinated expression and regulation of coassembly of the three subunits. Another report, using radioligand binding and co-immunoprecipitation studies in transfected *Drosophila* S2 cells, has suggested that Dβ3 can coassemble with Dα2, Dα3, Dα4 or Dβ2.[43]

Role for Dα7 in Drosophila Escape Behaviour

The powerful genetic toolkit available in the model organism, *D. melanogaster*, can yield insights into behavioural roles for individual nAChR subunits. This was demonstrated by a study which used immunohistochemistry to show that Dα7 protein is enriched in the dendrites of the giant fiber system which serves as a reflex circuit that triggers escape behaviours.[50] A fly strain with mutated Dα7 showed no obvious abnormalities when compared with wild-type flies but mutant adult flies did fail to exhibit the giant fiber-mediated startle response to a sudden change in light levels, indicating that Dα7 mediates the Drosophila escape response.

Uncovering the Actions of Imidacloprid and Spinosad Using Drosophila nAChRs

Work with Drosophila nAChRs has implicated certain subunits as targets of imidacloprid action. This has involved the use of heterologous expression systems such as *Xenopus laevis* (African clawed frog) oocytes[52] or a *D. melanogaster* cell line (Schneider S2 cells)[53] to study functional receptors. Unlike vertebrate nAChRs, reconstituting functional insect nAChRs in heterologous systems has proven elusive. Nevertheless, the fact that several Drosophila nAChR subunits can form functional nAChRs when co-expressed with a vertebrate β2 subunit in Xenopus oocytes has been exploited to identify Dα1 and Dα2 as candidate imidacloprid targets since Dα1/β2 and Dα2/β2 hybrid nAChRs were more neonicotinoid sensitive than the complete vertebrate α4/β2 receptor.[54,55] Also, the partial agonist actions of imidacloprid (and super-agonist actions of the second generation neonicotinoid clothianidin) reported for native Drosophila receptors are mimicked in the Dα2/β2 hybrid.[30] This approach has been extended to study whether vertebrate nAChRs (usually α4/β2 or α7) can be rendered more sensitive to neonicotinoids when insect nAChR-specific amino acids or

Table 2. Orders and key roles of insect species that have their complete nAChR gene family described

Species	Order	Importance	Genome Size (Mega Bases)	nAChR Subunit Gene Number
A. gambiae	Diptera	malaria vector	278	10
A. mellifera	Hymenoptera	pollination, honey production, social and behavioural model	262	11
B. mori	Lepidoptera	silk production, Lepidopteran model	429	12
D. melanogaster	Diptera	genetic model organism	118	10
T. castaneum	Coleoptera	pest of stored food, Coleopteran model	204	12

subunit regions have been introduced. In this way, residues in loops C,[56] D,[57-59] E[60] and F[58] as well as an insertion in loop F[56] have been shown to contribute to imidacloprid sensitivity. Also, Dβ1[57] and Dβ2[59] have been highlighted as additional subunits targeted by neonicotinoids. These studies using amino acid substitutions have led to the postulation that the formation of hydrogen bond networks plays a key role in neonicotinoid interactions.[61] Support for this view is also derived from structural studies in which snail AChBP bound with neonicotinoids has been crystallised.[62,63]

Spinosad is an insecticide which is derived from fermentation products of the soil dwelling bacterium *Saccharopolyspora spinosa.*[64] It acts on nAChRs but not at the same site as imidacloprid,[65] indicating they may act on separate nAChR types.[66] Indeed this is likely to be the case as a Dα6 knockout mutant strain of *D. melanogaster* was shown to be 1181-fold more resistant to spinosad than the control strain, identifying this subunit as a major spinosad target.[67]

Characterisation of Complete nAChR Gene Families from Five Insect Species Spanning Over 300 Million Years of Evolution

A Core Group of nAChR Subunits Is Highly Conserved in Different Insect Species

Since the publication of the *D. melanogaster* genome in 2000,[47] the genomes of several other insect species have since been sequenced. This information has so far been used to characterise the complete nAChR gene families from *Anopheles gambiae* (malaria mosquito),[68,69] *Apis mellifera* (honey bee),[70,71] *Tribolium castaneum* (red flour beetle)[72,73] and *Bombyx mori* (silk worm).[74,75] These species represent diverse orders which span over 300 million years of evolution (Table 2)[73] during which the nAChR gene families of these five insect species have remained compact consisting of 10 (*D. melanogaster* and *A. gambiae*), 11 (*A. mellifera*) or 12 (*B. mori* and *T. castaneum*) subunits.

Each of the five insect nAChR gene families has seven core groups of subunits that are highly conserved between species (Fig. 2).[76] Thus, Anopheles, Apis, Bombyx and Tribolium have subunit equivalents of Dα1-7, Dβ1 and Dβ2. The different insect species have the same number of core group subunits with the exception of *T. castaneum* which has an extra Dβ2-like subunit arising most likely through a gene duplication event (Fig. 2).[72] Dα5, Dα6 and Dα7 have been placed into a single group (Fig. 2) due to their considerable sequence homology with vertebrate α7 subunits (Table 3).[76,77] The presence of α7-like subunits also in nematodes[78,79] and trematodes[80] indicates an ancient lineage for this receptor subtype. The remaining insect

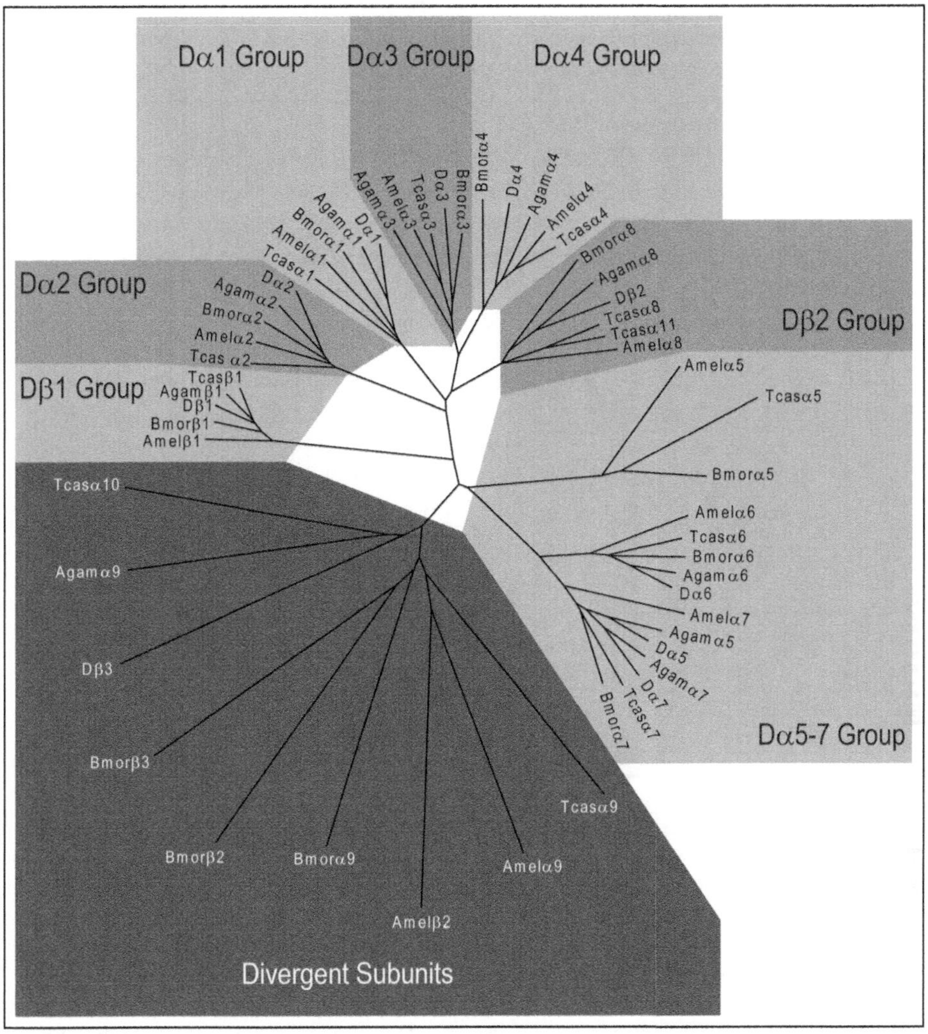

Figure 2. Tree showing the nAChR gene families *A. gambiae, A. mellifera, B. mori, D. melanogaster* and *T. castaneum*. Based on their high amino acid sequence homology, several insect nAChR subunits cluster into groups. Each insect possesses at least one divergent subunit that does not fall into any of these groups.

subunits do not have such close sequence relationships with those of vertebrates. Subunit homologs have also been found in other species such as *Ctenocephalides felis* (cat flea),[81] *Locusta migratoria* (migratory locust),[82] *Musca domestica* (house fly),[83-85] *Myzus persicae* (green peach aphid),[86,87] *Nilaparvata lugens* (brown plant hopper)[88] and *Schistocerca gregaria* (desert locust),[89,90] suggesting that the core groups are common to insects. Generally, equivalent nAChR subunits from different insect species have greater than 60% identity in their amino acid sequences. In addition to amino acid identity, distinct features are also conserved in core group nAChR subunits as summarised in Table 3. Interestingly, insect orthologs of Dβ2 are α subunits (e.g., Agamα8 and Amelα8 in Fig. 2), suggesting a change in functional role of the subunit in the Drosophila lineage.

Table 3. Amino acid sequence features particular to insect nAChR subunits when comparing Anopheles, Apis, Bombyx, Drosophila and Tribolium

Group	% Sequence Identity to Closest Human Homolog	Notable Features
Dα1	38-40% to human α2	Polypeptide insert in loop F which is involved in ligand binding
Dα2	36-38% to human α2	Polypeptide insert in loop F which is involved in ligand binding
Dα3	28-40% to human α2	Polypeptide insert in loop F which is involved in ligand binding. Agamα3 and Dα3 have unusually long intracellular domains between TM3 and TM4
Dα4	38-39% to human α2	Polypeptide insert in loop F which is involved in ligand binding. Alternative splicing of exon 4
Dα5-7	42-46% to human α7 Apis, Bombyx and Tribolium α5 have lower identity	Insect α6 subunits have alternative splicing of exons 3 and 8. Insect α6 subunits have conserved and distinct RNA A-to-I editing except for Agamα6
Dβ1	39-40% to human α2	
Dβ2	38-40% to human α2	Polypeptide insert in loop F which is involved in ligand binding. Is an α subunit in non-Drosophila species
Divergent	12-22% to human α2	Short intracellular domain between TM3 and TM4. Several divergent subunits lack the GEK amino acid motif preceding TM2 which is important for cation selectivity

Insect Species Possess a Distinct Complement of Divergent nAChR Subunits

Analysis of the five complete nAChR gene families has shown that insects possess at least one divergent subunit (Fig. 2) that shows low sequence homology to all other known nAChR subunits (less than 29% identity). Unlike core group subunits, analogous divergent subunits in different insects are difficult to assign. In addition to low sequence homology, divergent subunits possess extremely small intracellular domains between TM3 and TM4 and several examples, particularly those of *B. mori*, lack the highly conserved GEK amino acid motif preceding TM2[74] which is important for cation selectivity.[91] These subunits do not possess amino acid residues known to confer anion selectivity but they may form nAChRs with distinct ion channel characteristics. Currently, little is known about divergent nAChR function although it has been shown that Dβ3 can co-assemble with other nAChR subunits and influence ligand binding.[43] Each of the five insect species possesses a different set of divergent nAChR subunits. For example, *T. castaneum* has two divergent subunits which are both α,[72] *A. mellifera* also possesses two divergent subunits but one is α and the other β[70] whilst there are three divergent subunits (one α and two β) in *B. mori*.[74] Thus, the divergent subunits may perform species-specific roles and therefore be of interest as targets to control insect pests while sparing beneficial species.

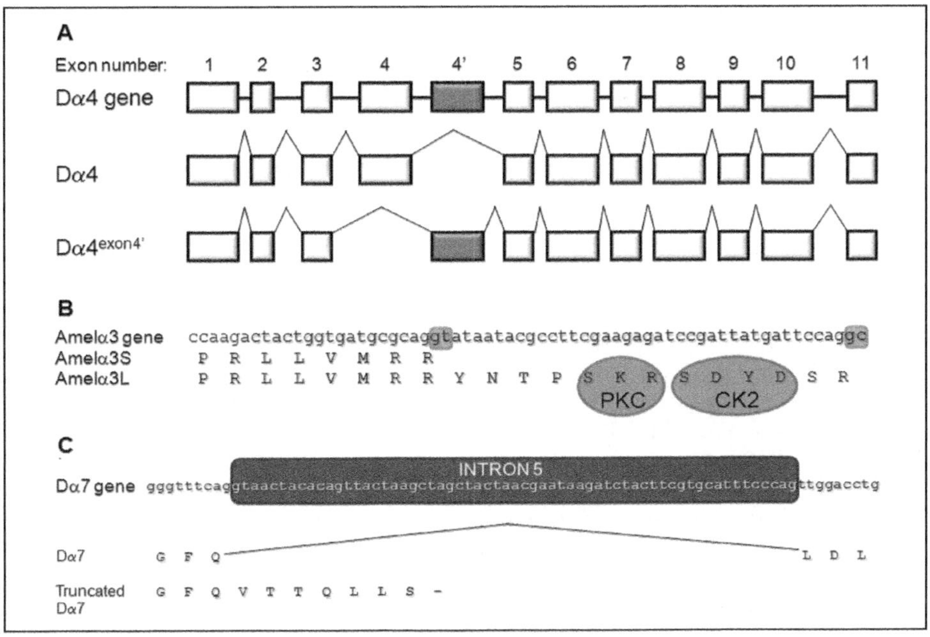

Figure 3. Examples of different forms of alternative splicing in insect nAChR subunits. A) Alternative splicing of exons. Dα4 possesses two alternatives for exon 4 (denoted exon 4 and exon 4')[42] which most likely arose through tandem exon duplication.[125] B) Use of different splice donor sites (highlighted in grey boxes) in the Amelα3 gene generates two intracellular domains differing in size by 13 amino acids.[70] The long variant (Amelα3L) has extra protein kinase C (PKC) and casein kinase 2 (CK2) phosphorylation sites which may regulate various aspects of receptor function.[94,95] C) A truncated variant of Dα7 is generated by the failure to excise intron 5 which introduces a premature stop codon (shown as a dash).[40]

RNA Editing and Alternative Splicing Broadens the Insect Nicotinic "Recepterome"

Alternative Splicing

Two Drosophila nAChR subunits (Dα4 and Dα6)[40,42] and their orthologs in other species have exons that are alternatively spliced, which effectively substitutes amino acids in regions important for receptor function and assembly.[69,70,72,74] For example, insect α4 subunits have two alternatives for exon 4 (denoted exon 4 and exon 4')[42] with different residues within, or in the vicinity of, the Cys-loop, which has been shown to be important for complete receptor assembly (Fig. 3).[4] Consistent with this, radioligand-binding assays indicate that Dα4 containing exon 4' assembles less efficiently than subunits with exon 4.[42] Interestingly, RT-PCR analysis revealed that the two Amelα4 splice variants are differentially expressed throughout the honey bee life cycle with exon 4 variants present at each developmental stage whereas exon 4' variants were detected only in pupae and adults.[70] This suggests that exon 4' subunits may serve to modulate receptor assembly in the later stages of honeybee development. Conservation of alternative splicing can also be seen in insect α6 subunits for exons 3 and 8[40,69,70,72,74] although the number of alternative exons can vary between species. For example, Agamα6 and Bmorα6 have two alternatives for exon 8[69,92] while Amelα6 and Dα6 have three.[40,92] Different residues introduced in functionally significant regions through alternative splicing can also vary between species. In one case, alternative splicing of Agamα6 exon 8 substitutes a valine for a leucine in the TM2 domain which lines the ion channel, whereas the leucine residue is conserved in all splice variants of Amelα6, Bmorα6, Dα6 and Tcasα6.[69,92] A mutation of the equivalent leucine in chick α7 to valine resulted in nAChRs with a reduced rate of desensitisation and an enhanced sensitivity to ACh.[93] Thus, alternative splicing may generate nAChR subunit isoforms with functional properties particular to certain insect species. Alternative splicing of exons can also be species specific. For example, Bmorα8 is the only known member of the Dβ2 group to have alternative splicing of exon 7, which introduces variation in TM2 and TM3 thereby potentially giving rise to variants with distinct ion channel properties.[74]

Species-specific nAChR subunit isoforms can also be generated through the differential use of splice sites, as in Amelα3 where two variants (long and short forms) have TM3-TM4 intracellular loops that differ in length by 13 amino acid residues (Fig. 3).[70] The long form has two extra putative phosphorylation sites which may have an effect on receptor properties since phosphorylation of the TM3-TM4 intracellular loop regulates several aspects of receptor function such as desensitisation and aggregation and could affect the action of insecticides.[77,94,95]

For several insect nAChR subunits, truncated transcripts have been detected where an exon is missing or where premature stop codons have been introduced either by omission of an exon which results in a frame shift or lack of splicing an intron (Fig. 3).[40,42,69,70,74,96] It remains to be determined whether the truncated transcripts are translated into proteins in vivo and if so it will be of interest to determine their role. It has been suggested that they may act as an 'ACh sponge' serving to terminate cholinergic transmission in a manner similar to that of the molluscan ACh-binding protein[40,96,97] although their ability to interact with ACh is questionable since all truncated subunits, with the exception of truncated Amelα3,[70] lack at least one loop involved in ligand binding. Another possible role is to regulate receptor expression similar to a truncated variant of the mouse α7 nAChR subunit which acts as a dominant negative when cotransfected with full length α7 in HEK 293 cells.[98]

RNA Editing

RNA A-to-I editing involves the modification of select adenosine (A) residues to inosine (I) in pre-mRNA transcripts by adenosine deaminases acting on RNA (ADARs).[99] Since inosine is interpreted by cellular machineries as guanosine (G), A-to-I editing generates transcripts with a nucleotide composition different from that of the corresponding genomic DNA (Fig. 4). This has the potential to alter amino acid residues thus generating multiple protein isoforms. RNA editing occurs particularly in gene products which are involved in neuronal signaling,[100] consistent with neurological phenotypes observed for ADAR-deficient *Caenorhabditis elegans*,[101] *Drosophila melanogaster*[102] and mice.[103]

RNA A-to-I editing has been observed in five *D. melanogaster* nAChR subunits which alter amino acid residues in functionally significant regions.[40,50,100] For example, editing of Dα5, Dα7 and Dβ2 alters residues in the TM2, 3 and 4 domains, thereby potentially affecting ion channel characteristics.[6,104,105] RNA editing may also affect the ligand binding properties of two subunits (Dα6 and Dβ1) since residues are altered in the extracellular N-terminal region. RNA editing is less widespread in nAChR subunits of other insect species. For instance, two *T. castaneum* nAChR subunits (Tcasα6 and Tcasβ1) are edited[72] whilst in *A. mellifera* RNA editing was only seen in Amelα6.[70] RNA editing of Amelα6 alters nine amino acid residues in a confined area located in the vicinity of loop E.[70,92] Up to five of these residues are also altered through editing in other insects such as *B. mori*, *D. melanogaster*, *H. viriscens*, *M. domestica* and *T. castaneum* (Fig. 4).[85,92] Interestingly, conserved editing in α6 of different species removes an N-glycosylation site in loop E which may affect receptor maturation, channel desensitisation and conductance.[106,107] The reverse appears to be the case for *M. domestica* where the equivalent N-glycosylation site is created through editing of asparagine to serine (Fig. 4).[85] Several editing sites in Amelα6, however, are not conserved in other insects and no RNA editing at all was

A
Transcription
Splicing
Translation
AAGATAGAC
Genomic cDNA
Edited pre-mRNA
Non-edited pre-mRNA
AAGATIGAC
AAGATAGAC
Intron
Exon
AAG|ATG|GAC
AAG|ATA|GAC
K M D
Polypeptide recoded by RNA editing
K I D
Polypeptide encoded by genomic DNA
B
LpE
LpB
LpF
Ameloα6 SADEGFDGTYQTNVVVTHNGSCLYVPPGIFKSTCKIDITWFPFDDQHCDMKFGSWTYDGNQVDLVLSSET
Agamα6 SADEGFDGTYHTNIVVKNNGSCLYVPPGIFKSTCKIDITWFPFDDQHCEMKFGSWTYDGNQLDLILNSDD
Bmorα6 SADEGFDSTYPTNVVVRNNGSCLYVPPGIFKSTCKIDITWFPFDDQRCEMKFGSWTYDGYQLDLQLQDEA
Dα6 SADEGFDGTYHTNIVVKHNGSCLYVPPGIFKSTCKIDITWFPFDDQHCEMKFGSWTYDGNQLDLVLNSED
Hvα7-2 SADEGFDGTYQTNVVVRHNGSCLYVPPGIFKSTCKIDIAWFPFDDQHCDMKFGSWTYDGNQLDLVLKDEA
Mdα6 SADEGFDGTYHTNIVVKHNGNCLYVPPGIFKSTCKIDITWFPFDDQHCEMKFGSWTYDGNQLDLVLNSED
Tcasα6 SADEGFDGTFQTNVVVKHNGSCLYVPPGIFKSTCKIDITWFPFDDQHCDMKFGSWTYDGNQLDLVLNSES

Figure 4. RNA A-to-I editing recodes the genome. A) Schematic of the RNA editing process. Select adenosine (A) residues in pre-mRNA are modified to inosine (I) by adenosine deaminases acting on RNA (ADARs). Since inosine is interpreted by cellular machineries as guanosine (G), A-to-I editing generates mRNA transcripts with a nucleotide composition that differs from the corresponding genomic DNA. This has the potential to alter amino acid residues thus generating multiple protein isoforms. B) A comparison of RNA A-to-I editing in Dα6 with orthologs in *A. gambiae*, *A. mellifera*, *B. mori*, *H. virescens*, *M. domestica* and *T. castaneum*. Ligand binding domains LpB, LpE and LpF are indicated and amino acids altered by RNA editing are highlighted in white text.

detected in Agamα6 of *A. gambiae*.[69] Thus, RNA editing can generate species-specific nAChR subunit isoforms. It has been observed that genomically-encoded guanosines in certain insect α6 subunits including Agamα6 are in fact A-to-I editing sites in other species leading to the suggestion that RNA editing maintains phylogenetic conservation while broadening protein diversity possibly as part of an evolutionary mechanism.[92,108] Studies on Drosophila have shown that RNA editing is particularly important in the nervous system function of adults.[102] This may hold true for other insect species since the greatest extent of nAChR RNA editing was observed in adults of *A. mellifera*.[70]

Conclusion and Prospects

Small Gene Families with Large-Scale Proteome Diversity

Characterisation of the first two complete insect nAChR gene families, those from the diptera *D. melanogaster* and *A. gambiae*, revealed a small complement of subunits numbering 10 in both species.[48,69,96] It was surprising that the third insect nAChR gene family to be described, that of the hymenoptera *A. mellifera*, possessed a similar number of subunits (11)[70] since the honey bee displays a far more complex behavioural repertoire than either the fruit fly or malaria mosquito. With the characterisation of *B. mori*[74] and *T. castaneum*[72] nAChR subunits, a consensus emerged that insect nAChR gene families remained compact over 300 million years of evolution. However, while the gene numbers are relatively small compared to other organisms, the number of insect nAChR gene products can be much larger due to alternative splicing and RNA editing which have the potential to generate a receptor proteome with diversity far greater than that suggested by the number of genes alone. In addition to broadening the nAChR proteome in a given insect species, alternative splicing and RNA editing generates species-specific subunit variants with potentially distinct functional characteristics. Since subunit composition determines nAChR pharmacological and functional characteristics, a major goal in determining proteome diversity would be to elucidate the stoichiometry of subunits and their isoforms in insect nAChRs in vivo.

Upcoming New nAChR Gene Families of Interest

Genome projects have either been completed or are in progress enabling the characterisation of nAChR gene families from other insect species. For example, the genome sequences of 12 Drosophila species were published in 2007[109] providing a far greater scope for comparative genome data analysis and studying with fine resolution nAChR diversity in a single phyla. The yellow and dengue fever mosquito *Aedes aegypti* genome has also been published[110] allowing for comparative studies with the malaria mosquito as well as with other species. Genome projects currently underway include those of the West Nile virus mosquito *Culex pipiens* (http://www.broad.mit.edu/annotation/genome/culex_pipiens.4/Info.html), the human body louse *Pediculus humanus humanus*,[111] the pea aphid *Acyrthosiphon pisum* (http://www.hgsc.bcm.tmc.edu/projects/aphid/) which is an agricultural pest and the parasitoid wasp *Nasonia vitripennis* (http://www.hgsc.bcm.tmc.edu/projects/nasonia/) which is an important organism in the biological control of insect pests. Sequence information produced from such projects will provide further insights into the diversity of insect nAChR gene families. Cross hybridisation approaches still nevertheless have an important role to play in determining nAChR sequences of insects for which no genome information is currently available. For example, the cockroach *Periplaneta americana* played an important role as an early insect neurobiology model providing access to an identified cholinergic synapse and thereby facilitating combined biochemical and electrophysiological studies.[112-114] This orthopteran species is now being explored by Lapied and colleagues to determine the members of the nAChR family and their functional roles.

Behavioural Studies, Forward and Reverse Genetics in Dissecting Functional Roles of nAChRs

Genome sequence information and well-characterised nAChR gene families provide an invaluable basis for the further study of nAChR functional diversity. As we have illustrated the study of *D. melanogaster* mutants has the potential to pinpoint single nAChR subunits either in particular behavioural roles or as insecticide targets. These are examples of forward genetics which aim to find the genetic basis of a phenotype or trait. *B. mori* is the second most widely used genetic model insect after *D. melanogaster* due to the ease of their rearing and the availability of mutants from genetically homogenous inbred lines serve as a potentially useful tool for forward genetic studies.[75]

Reverse genetics, as the name implies, proceeds in the opposite direction of forward genetics by seeking to determine possible phenotypes arising from a specific DNA sequence. This is usually achieved by knocking down the function of a gene of interest. An example involves creating *D. melanogaster* mutants for Dα7 by using P-elements to assess the role of the subunit in vivo.[50] RNA interference (RNAi) is a powerful reverse genetics approach, first characterised in *C. elegans*, involving the introduction of double-stranded RNA which results in silencing of the corresponding gene.[115] In 2003, a genome-wide RNAi screen was published using *C. elegans* which involved silencing 16,757 genes (corresponding to approximately 86% of the genome) in a general survey of gene function.[116] More recently, genome-wide screens covering over 90% of the *D. melanogaster* genome have been applied to Drosophila cells to identify genes playing roles in specific processes, one example being neural outgrowth.[117] No nAChR subunits were implicated in this study but a similar screen specifically addressing cholinergic signalling may reveal the importance of various nAChR subunits as well as identify novel genes involved in nAChR signalling. Parental RNAi, where RNA interference arising from double-stranded RNA introduced into the mother also spreads to the offspring, is highly efficient in *T. castaneum*.[73] Thus, the beetle provides a powerful tool for studying nAChR gene function in an insect pest species.

A. mellifera is a key model for social behaviour as well as learning and these features have been exploited in studies of the involvement of nAChRs in honey bee behaviour. Injection of the nAChR agonist, nicotine, showed that potentiation of the cholinergic system improves short-term memory[118] and injection of the nAChR antagonist, mecamylamine, inhibited olfactory learning or memory recall depending on the site of injection.[119,120] It has also been demonstrated that one distinct nAChR subtype, which is sensitive to the antagonist α-bungarotoxin, is involved in long-term memory, whereas a second subtype, which is insensitive to α-bungarotoxin but is affected by mecamylamine, plays a role in retrieving information stored during single-trial learning.[121] Interestingly, this mirrors to a certain extent the mammalian central nervous system where there are two predominant nAChR subtypes, α7 and α4/β2 receptors, that are α-bungarotoxin sensitive and insensitive, respectively, both of which play a role in memory.[122] The development of compounds known to target specific honey bee nAChRs will allow these behavioural studies to be performed with finer resolution to elucidate the role of particular subunits in various aspects of behaviour.

Towards a New Era of Improved, Safer Pesticide Design

The characterisation of complete insect nAChR gene families has shown that while it is evident that most nAChR subunits are highly conserved between diverse insect species, alternative splicing and RNA editing as well as the presence of divergent subunits present species-specific isoforms which can perhaps be exploited for the development of compounds that target particular insects pests such as *A. gambiae* and *T. castaneum* while sparing beneficial insects such as *A. mellifera* and *B. mori*. Computer three-dimensional models of insect nAChRs have been generated based on the snail AChBP which permit docking experiments to assess interactions with compounds of interest.[96] Also, the *T. marmorata* nAChR X-ray structure was used to build models of five theoretical subtypes of *A. mellifera* nAChRs (α1/β1, α3/β2, α4/β2, α6/β2 and α9).[123] Docking simulations showed that both imidacloprid and the insecticide fipronil, which blocks GABA-gated chloride channels, bind to the honey bee nAChRs with the involvement of numerous hydrogen bonds and hydrophobic interactions, the number of which varied depending on receptor subtype. Now that crystal structures are available for AChBP with imidacloprid and other neonicotinoids docked,[62,63] further improvements of such models can be anticipated.

A major goal yet to be achieved which would greatly facilitate the search for improved/novel insecticides is the successful expression of functional insect nAChRs in heterologous systems such as *Xenopus laevis* oocytes or cell lines. This would enable testing of numerous compounds on nAChRs of known subunit composition. Together with molecular modelling, this would likely prove invaluable in screening for compounds that show selectivity for specific nAChR subtypes, thereby enhancing safety and providing guidelines for minimising adverse effects on beneficial species, as well as facilitating an improved understanding of insecticide-receptor interactions.

Acknowledgements

The authors are indebted to The Medical Research Council of the UK for support.

References

1. Sine SM, Engel AG. Recent advances in Cys-loop receptor structure and function. Nature 2006; 440:448-455.
2. Connolly CN, Wafford KA. The Cys-loop superfamily of ligand-gated ion channels: the impact of receptor structure on function. Biochem Soc Trans. 2004; 32:529-534.
3. Changeux JP, Taly A. Nicotinic receptors, allosteric proteins and medicine. Trends Mol Med. 2008; 14:93-102.
4. Green WN, Wanamaker CP. The role of the cystine loop in acetylcholine receptor assembly. J Biol Chem. 1997; 272:20945-20953.
5. Grutter T, de Carvalho LP, Dufresne V et al. Molecular tuning of fast gating in pentameric ligand-gated ion channels. Proc Natl Acad Sci. USA 2005; 102:18207-18212.
6. Corringer PJ, Le Novere N, Changeux JP. Nicotinic receptors at the amino acid level. Annu Rev Pharmacol Toxicol. 2000; 40:431-458.
7. Kao PN, Karlin A. Acetylcholine receptor binding site contains a disulfide cross-link between adjacent half-cystinyl residues. J Biol Chem. 1986; 261:8085-8088.
8. Millar NS, Gotti C. Diversity of vertebrate nicotinic acetylcholine receptors. Neuropharmacology. 2009;56(1):237-46
9. Unwin N. Refined structure of the nicotinic acetylcholine receptor at 4A resolution. J Mol Biol. 2005; 346:967-989.
10. Dellisanti CD, Yao Y, Stroud JC et al. Crystal structure of the extracellular domain of nAChR alpha1 bound to alpha-bungarotoxin at 1.94 A resolution. Nat Neurosci. 2007; 10:953-962.
11. Changeux JP, Kasai M, Lee CY. Use of a snake venom toxin to characterize the cholinergic receptor protein. Proc Natl Acad Sci USA. 1970; 67:1241-1247.
12. Brejc K, van Dijk WJ, Klaassen RV et al. Crystal structure of an ACh-binding protein reveals the ligand-binding domain of nicotinic receptors. Nature 2001; 411:269-276.
13. Hilf RJ, Dutzler R. X-ray structure of a prokaryotic pentameric ligand-gated ion channel. Nature 2008; 452:375-379.
14. Hogg RC, Bertrand D. Nicotinic acetylcholine receptors as drug targets. Curr Drug Targets CNS Neurol Disord. 2004; 3:123-130.
15. Combi R, Dalpra L, Tenchini ML et al. Autosomal dominant nocturnal frontal lobe epilepsy—a critical overview. J Neurol. 2004; 251:923-934.
16. Engel AG, Sine SM. Current understanding of congenital myasthenic syndromes. Curr Opin Pharmacol. 2005; 5:308-321.
17. Hughes BW, Moro De Casillas ML, Kaminski HJ. Pathophysiology of myasthenia gravis. Semin Neurol. 2004; 24:21-30.

18. Watson R, Jepson JE, Bermudez I et al. Alpha7-acetylcholine receptor antibodies in two patients with Rasmussen encephalitis. Neurology 2005; 65:1802-1804.
19. Gotti C, Zoli M, Clementi F. Brain nicotinic acetylcholine receptors: native subtypes and their relevance. Trends Pharmacol Sci. 2006; 27:482-491.
20. Brown LA, Jones AK, Buckingham SD et al. Contributions from Caenorhabditis elegans functional genetics to antiparasitic drug target identification and validation: nicotinic acetylcholine receptors, a case study. Int J Parasitol. 2006; 36:617-624.
21. Kaminsky R, Ducray P, Jung M et al. A new class of anthelmintics effective against drug-resistant nematodes. Nature 2008; 452:176-180.
22. Baylis HA, Sattelle DB, Lane NJ. Genetic analysis of cholinergic nerve terminal function in invertebrates. J Neurocytol. 1996; 25:747-762.
23. Matsuda K, Buckingham SD, Kleier D et al. Neonicotinoids: insecticides acting on insect nicotinic acetylcholine receptors. Trends Pharmacol Sci. 2001; 22:573-580.
24. Jeschke P, Nauen R. Neonicotinoids-from zero to hero in insecticide chemistry. Pest Manag Sci. 2008; 64(11):1084-98.
25. Lind RJ, Clough MS, Reynolds SE et al. H-3 imidacloprid labels high- and low-affinity nicotinic acetylcholine receptor-like binding sites in the aphid Myzus persicae (Hemiptera: Aphididae). Pesticide Bioch Physiol. 1998; 62:3-14.
26. Liu M-Y, Casida JE. High-affinity binding of H-3 imidacloprid in the insect acetylcholine receptor. Pesticide Biochem Physiol. 1993; 46:40-46.
27. Zhang A, Kayser H, Maienfisch P et al. Insect nicotinic acetylcholine receptor: conserved neonicotinoid specificity of [(3)H]imidacloprid binding site. J Neurochem. 2000; 75:1294-1303.
28. Bai B, Lummis SCR, Leicht W et al. Actions of imidacloprid and a related nitromethylene on cholinergic receptors of an identified insect motor-neuron. Pescticide Sci. 1991; 33:197-204.
29. Jepson JE, Brown LA, Sattelle DB. The actions of the neonicotinoid imidacloprid on cholinergic neurons of Drosophila melanogaster. Invert Neurosci. 2006; 6:33-40.
30. Brown LA, Ihara M, Buckingham SD et al. Neonicotinoid insecticides display partial and super agonist actions on native insect nicotinic acetylcholine receptors. J Neurochem. 2006; 99:608-615.
31. Tomizawa M, Casida JE. Selective toxicity of neonicotinoids attributable to specificity of insect and mammalian nicotinic receptors. Annu Rev Entomol. 2003; 48:339-364.
32. Nishiwaki H, Nakagawa Y, Kuwamura M et al. Correlations of the electrophysiological activity of neonicotinoids with their binding and insecticidal activities. Pest Manag Sci 2003; 59:1023-1030.
33. Ballivet M, Patrick J, Lee J et al. Molecular cloning of cDNA coding for the gamma subunit of Torpedo acetylcholine receptor. Proc Natl Acad Sci USA. 1982; 79:4466-4470.
34. Devillers-Thiery A, Giraudat J, Bentaboulet M et al. Complete mRNA coding sequence of the acetylcholine binding alpha-subunit of Torpedo marmorata acetylcholine receptor: a model for the transmembrane organization of the polypeptide chain. Proc Natl Acad Sci USA. 1983; 80:2067-2071.
35. Noda M, Takahashi H, Tanabe T et al. Primary structure of alpha-subunit precursor of Torpedo californica acetylcholine receptor deduced from cDNA sequence. Nature 1982; 299:793-797.
36. Noda M, Takahashi H, Tanabe T et al. Primary structures of beta- and delta-subunit precursors of Torpedo californica acetylcholine receptor deduced from cDNA sequences. Nature 1983; 301:251-255.
37. Sumikawa K, Houghton M, Smith JC et al. The molecular cloning and characterisation of cDNA coding for the alpha subunit of the acetylcholine receptor. Nucleic Acids Res. 1982; 10:5809-5822.
38. Hermans-Borgmeyer I, Zopf D, Ryseck RP et al. Primary structure of a developmentally regulated nicotinic acetylcholine receptor protein from Drosophila. EMBO J. 1986; 5:1503-1508.
39. Bossy B, Ballivet M, Spierer P. Conservation of neural nicotinic acetylcholine receptors from Drosophila to vertebrate central nervous systems. EMBO J. 1988; 7:611-618.
40. Grauso M, Reenan RA, Culetto E et al. Novel putative nicotinic acetylcholine receptor subunit genes, Dalpha5, Dalpha6 and Dalpha7, in Drosophila melanogaster identify a new and highly conserved target of adenosine deaminase acting on RNA-mediated A-to-I pre-mRNA editing. Genetics 2002; 160:1519-1533.
41. Jonas P, Baumann A, Merz B et al. Structure and developmental expression of the D alpha 2 gene encoding a novel nicotinic acetylcholine receptor protein of Drosophila melanogaster. FEBS Lett. 1990; 269:264-268.
42. Lansdell SJ, Millar NS. Cloning and heterologous expression of Dalpha4, a Drosophila neuronal nicotinic acetylcholine receptor subunit: identification of an alternative exon influencing the efficiency of subunit assembly. Neuropharmacology 2000; 39:2604-2614.

43. Lansdell SJ, Millar NS. Dbeta3, an atypical nicotinic acetylcholine receptor subunit from Drosophila: molecular cloning, heterologous expression and coassembly. J Neurochem. 2002; 80:1009-1018.
44. Sawruk E, Schloss P, Betz H et al. Heterogeneity of Drosophila nicotinic acetylcholine receptors: SAD, a novel developmentally regulated alpha-subunit. EMBO J. 1990; 9:2671-2677.
45. Sawruk E, Udri C, Betz H et al. SBD, a novel structural subunit of the Drosophila nicotinic acetylcholine receptor, shares its genomic localization with two alpha-subunits. FEBS Lett. 1990; 273:177-181.
46. Schulz R, Sawruk E, Mulhardt C et al. D alpha3, a new functional alpha subunit of nicotinic acetylcholine receptors from Drosophila. J Neurochem. 1998; 71:853-862.
47. Adams MD, Celniker SE, Holt RA et al. The genome sequence of Drosophila melanogaster. Science 2000; 287:2185-2195.
48. Littleton JT, Ganetzky B. Ion channels and synaptic organization: analysis of the Drosophila genome. Neuron 2000; 26:35-43.
49. Jones AK, Davis P, Hodgkin J et al. The nicotinic acetylcholine receptor gene family of the nematode Caenorhabditis elegans: an update on nomenclature. Invert Neurosci. 2007; 7:129-131.
50. Fayyazuddin A, Zaheer MA, Hiesinger PR et al. The nicotinic acetylcholine receptor Dalpha7 is required for an escape behavior in Drosophila. PLoS Biol. 2006; 4:e63.
51. Chamaon K, Smalla KH, Thomas U et al. Nicotinic acetylcholine receptors of Drosophila: three subunits encoded by genomically linked genes can co-assemble into the same receptor complex. J Neurochem. 2002; 80:149-157.
52. Buckingham SD, Pym L, Sattelle DB. Oocytes as an expression system for studying receptor/channel targets of drugs and pesticides. Methods Mol Biol. 2006; 322:331-345.
53. Towers PR, Sattelle DB. A Drosophila melanogaster cell line (S2) facilitates postgenome functional analysis of receptors and ion channels. Bioessays 2002; 24:1066-1073.
54. Ihara M, Matsuda K, Otake M et al. Diverse actions of neonicotinoids on chicken alpha7, alpha4beta2 and Drosophila-chicken SADbeta2 and ALSbeta2 hybrid nicotinic acetylcholine receptors expressed in Xenopus laevis oocytes. Neuropharmacology 2003; 45:133-144.
55. Matsuda K, Buckingham SD, Freeman JC et al. Effects of the alpha subunit on imidacloprid sensitivity of recombinant nicotinic acetylcholine receptors. Br J Pharmacol. 1998; 123:518-524.
56. Shimomura M, Yokota M, Matsuda K et al. Roles of loop C and the loop B-C interval of the nicotinic receptor alpha subunit in its selective interactions with imidacloprid in insects. Neurosci Lett. 2004; 363:195-198.
57. Shimomura M, Yokota M, Ihara M et al. Role in the selectivity of neonicotinoids of insect-specific basic residues in loop D of the nicotinic acetylcholine receptor agonist binding site. Mol Pharmacol. 2006; 70:1255-1263.
58. Shimomura M, Yokota M, Okumura M et al. Combinatorial mutations in loops D and F strongly influence responses of the alpha7 nicotinic acetylcholine receptor to imidacloprid. Brain Res. 2003; 991:71-77.
59. Shimomura M, Okuda H, Matsuda K et al. Effects of mutations of a glutamine residue in loop D of the alpha7 nicotinic acetylcholine receptor on agonist profiles for neonicotinoid insecticides and related ligands. Br J Pharmacol. 2002; 137:162-169.
60. Amiri S, Shimomura M, Vijayan R et al. A role for Leu118 of loop E in agonist binding to the alpha 7 nicotinic acetylcholine receptor. Mol Pharmacol. 2008; 73:1659-1667.
61. Ihara M, Shimomura M, Ishida C et al. A hypothesis to account for the selective and diverse actions of neonicotinoid insecticides at their molecular targets, nicotinic acetylcholine receptors: catch and release in hydrogen bond networks. Invert Neurosci. 2007; 7:47-51.
62. Ihara M, Okajima T, Yamashita A et al. Crystal structures of Lymnaea stagnalis AChBP in complex with neonicotinoid insecticides imidacloprid and clothianidin. Invert Neurosci. 2008; 8:71-81.
63. Tomizawa M, Maltby D, Talley TT et al. Atypical nicotinic agonist bound conformations conferring subtype selectivity. Proc Natl Acad Sci USA. 2008; 105:1728-1732.
64. Bret BL. Biological properties of spinosad. Down Earth 1997; 52:6-13.
65. Salgado VL. The modes of action of spinosad and other insect control products. Down Earth 1997; 52:35-43.
66. Salgado VL. Studies on the mode of action of spinosad: insect symptoms and physiological correlates. Pestic Biochem Physiol. 1998; 60:91-102.
67. Perry T, McKenzie JA, Batterham P. A Dalpha6 knockout strain of Drosophila melanogaster confers a high level of resistance to spinosad. Insect Biochem Mol Biol. 2007; 37:184-188.
68. Holt RA, Subramanian GM, Halpern A et al. The genome sequence of the malaria mosquito Anopheles gambiae. Science. 2002; 298:129-149.

69. Jones AK, Grauso M, Sattelle DB. The nicotinic acetylcholine receptor gene family of the malaria mosquito, Anopheles gambiae. Genomics. 2005; 85:176-187.
70. Jones AK, Raymond-Delpech V, Thany SH et al. The nicotinic acetylcholine receptor gene family of the honey bee, Apis mellifera. Genome Res. 2006; 16:1422-1430.
71. The Honeybee Genome Sequencing Consortium. Insights into social insects from the genome of the honeybee Apis mellifera. Nature. 2006; 443:931-949.
72. Jones AK, Sattelle DB. The cys-loop ligand-gated ion channel gene superfamily of the red flour beetle, Tribolium castaneum. BMC Genomics. 2007; 8:327.
73. Richards S, Gibbs RA, Weinstock GM et al. The genome of the model beetle and pest Tribolium castaneum. Nature. 2008; 452:949-955.
74. Shao YM, Dong K, Zhang CX. The nicotinic acetylcholine receptor gene family of the silkworm, Bombyx mori. BMC Genomics. 2007; 8:324.
75. Xia Q, Zhou Z, Lu C et al. A draft sequence for the genome of the domesticated silkworm (Bombyx mori). Science. 2004; 306:1937-1940.
76. Jones AK, Brown LA, Sattelle DB. Insect nicotinic acetylcholine receptor gene families: from genetic model organism to vector, pest and beneficial species. Invert Neurosci. 2007; 7(1):67-73.
77. Thany SH, Lenaers G, Raymond-Delpech V et al. Exploring the pharmacological properties of insect nicotinic acetylcholine receptors. Trends Pharmacol Sci. 2007; 28:14-22.
78. Mongan NP, Jones AK, Smith GR et al. Novel alpha7-like nicotinic acetylcholine receptor subunits in the nematode Caenorhabditis elegans. Protein Sci. 2002; 11:1162-1171.
79. Williamson SM, Walsh TK, Wolstenholme AJ. The cys-loop ligand-gated ion channel gene family of Brugia malayi and Trichinella spiralis: a comparison with Caenorhabditis elegans. Invert Neurosci. 2007; 7:219-226.
80. Bentley GN, Jones AK, Oliveros Parra WG et al. ShAR1alpha and ShAR1beta: novel putative nicotinic acetylcholine receptor subunits from the platyhelminth blood fluke Schistosoma. Gene. 2004; 329:27-38.
81. Bass C, Lansdell SJ, Millar NS et al. Molecular characterisation of nicotinic acetylcholine receptor subunits from the cat flea, Ctenocephalides felis (Siphonaptera: Pulicidae). Insect Biochem Mol Biol. 2006; 36:86-96.
82. Hermsen B, Stetzer E, Thees R et al. Neuronal nicotinic receptors in the locust Locusta migratoria. Cloning and expression. J Biol Chem. 1998; 273:18394-18404.
83. Gao JR, Deacutis JM, Scott JG. Characterization of the nicotinic acetylcholine receptor subunit gene Mdalpha2 from the house fly, Musca domestica. Arch Insect Biochem Physiol. 2007; 64:30-42.
84. Gao JR, Deacutis JM, Scott JG. The nicotinic acetylcholine receptor subunits Mdalpha5 and Mdbeta3 on autosome 1 of Musca domestica are not involved in spinosad resistance. Insect Mol Biol. 2007; 16:691-701.
85. Gao JR, Deacutis JM, Scott JG. The nicotinic acetylcholine receptor subunit Mdalpha6 from Musca domestica is diversified via posttranscriptional modification. Insect Mol Biol. 2007; 16:325-334.
86. Huang Y, Williamson MS, Devonshire AL et al. Cloning, heterologous expression and co-assembly of Mpbeta1, a nicotinic acetylcholine receptor subunit from the aphid Myzus persicae. Neurosci Lett. 2000; 284:116-120.
87. Sgard F, Fraser SP, Katkowska MJ et al. Cloning and functional characterisation of two novel nicotinic acetylcholine receptor alpha subunits from the insect pest Myzus persicae. J Neurochem. 1998; 71:903-912.
88. Liu Z, Williamson MS, Lansdell SJ et al. A nicotinic acetylcholine receptor mutation conferring target-site resistance to imidacloprid in Nilaparvata lugens (brown planthopper). Proc Natl Acad Sci USA. 2005; 102:8420-8425.
89. Jones AK, Marshall J, Blake AD et al. Sgbeta1, a novel locust (Schistocerca gregaria) non-alpha nicotinic acetylcholine receptor-like subunit with homology to the Drosophila melanogaster Dbeta1 subunit. Invert Neurosci. 2005; 5:147-155.
90. Marshall J, Buckingham SD, Shingai R et al. Sequence and functional expression of a single alpha subunit of an insect nicotinic acetylcholine receptor. EMBO J. 1990; 9:4391-4398.
91. Jensen ML, Schousboe A, Ahring PK. Charge selectivity of the Cys-loop family of ligand-gated ion channels. J Neurochem. 2005; 92:217-225.
92. Jin Y, Tian N, Cao J et al. RNA editing and alternative splicing of the insect nAChR subunit alpha6 transcript: evolutionary conservation, divergence and regulation. BMC Evol Biol. 2007; 7:98.
93. Revah F, Bertrand D, Galzi JL et al. Mutations in the channel domain alter desensitization of a neuronal nicotinic receptor. Nature 1991; 353:846-849.

94. Borges LS, Ferns M. Agrin-induced phosphorylation of the acetylcholine receptor regulates cytoskeletal anchoring and clustering. J Cell Biol. 2001; 153:1-12.
95. Hopfield JF, Tank DW, Greengard P et al. Functional modulation of the nicotinic acetylcholine receptor by tyrosine phosphorylation. Nature. 1988; 336:677-680.
96. Sattelle DB, Jones AK, Sattelle BM et al. Edit, cut and paste in the nicotinic acetylcholine receptor gene family of Drosophila melanogaster. Bioessays. 2005; 27:366-376.
97. Smit AB, Brejc K, Syed N et al. Structure and function of AChBP, homologue of the ligand-binding domain of the nicotinic acetylcholine receptor. Ann N Y Acad Sci. 2003; 998:81-92.
98. Saragoza PA, Modir JG, Goel N et al. Identification of an alternatively processed nicotinic receptor alpha7 subunit RNA in mouse brain. Brain Res Mol Brain Res 2003; 117:15-26.
99. Seeburg PH. A-to-I editing: new and old sites, functions and speculations. Neuron 2002; 35:17-20.
100. Hoopengardner B, Bhalla T, Staber C et al. Nervous system targets of RNA editing identified by comparative genomics. Science 2003; 301:832-836.
101. Tonkin LA, Saccomanno L, Morse DP et al. RNA editing by ADARs is important for normal behavior in Caenorhabditis elegans. EMBO J 2002; 21:6025-6035.
102. Palladino MJ, Keegan LP, O'Connell MA et al. A-to-I pre-mRNA editing in Drosophila is primarily involved in adult nervous system function and integrity. Cell 2000; 102:437-449.
103. Higuchi M, Maas S, Single FN et al. Point mutation in an AMPA receptor gene rescues lethality in mice deficient in the RNA-editing enzyme ADAR2. Nature 2000; 406:78-81.
104. Guzman GR, Santiago J, Ricardo A et al. Tryptophan scanning mutagenesis in the alphaM3 transmembrane domain of the Torpedo californica acetylcholine receptor: functional and structural implications. Biochemistry 2003; 42:12243-12250.
105. Tamamizu S, Lee Y, Hung B et al. Alteration in ion channel function of mouse nicotinic acetylcholine receptor by mutations in the M4 transmembrane domain. J Membr Biol 1999; 170:157-164.
106. Gehle VM, Walcott EC, Nishizaki T et al. N-glycosylation at the conserved sites ensures the expression of properly folded functional ACh receptors. Brain Res Mol Brain Res 1997; 45:219-229.
107. Nishizaki T. N-glycosylation sites on the nicotinic ACh receptor subunits regulate receptor channel desensitization and conductance. Brain Res Mol Brain Res 2003; 114:172-176.
108. Yang Y, Lv J, Gui B et al. A-to-I RNA editing alters less-conserved residues of highly conserved coding regions: implications for dual functions in evolution. RNA 2008; 14:1516-1525.
109. Clark AG, Eisen MB, Smith DR et al. Evolution of genes and genomes on the Drosophila phylogeny. Nature 2007; 450:203-218.
110. Nene V, Wortman JR, Lawson D et al. Genome sequence of Aedes aegypti, a major arbovirus vector. Science 2007; 316:1718-1723.
111. Pittendrigh BR, Clark JM, Johnston JS et al. Sequencing of a new target genome: the Pediculus humanus humanus (Phthiraptera: Pediculidae) genome project. J Med Entomol 2006; 43:1103-1111.
112. Harrow ID, Hue B, Pelhate M et al. Cockroach giant interneurones stained by cobalt-backfilling of dissected axons. J Exp Biol 1980; 84:341-343.
113. Lane NJ, Sattelle DB, Hufnagel LA. Pre- and postsynaptic structures in insect CNS: intramembranous features and sites of alpha-bungarotoxin binding. Tissue Cell 1983; 15:921-937.
114. Lummis SC, Sattelle DB. Binding of N-[propionyl-3H]propionylated alpha-bungarotoxin and L-[benzilic-4,4'-3H] quinuclidinyl benzilate to CNS extracts of the cockroach Periplaneta americana. Comp Biochem Physiol C 1985; 80:75-83.
115. Fire A, Xu S, Montgomery MK et al. Potent and specific genetic interference by double-stranded RNA in Caenorhabditis elegans. Nature 1998; 391:806-811.
116. Kamath RS, Fraser AG, Dong Y et al. Systematic functional analysis of the Caenorhabditis elegans genome using RNAi. Nature 2003; 421:231-237.
117. Sepp KJ, Hong P, Lizarraga SB et al. Identification of neural outgrowth genes using genome-wide RNAi. PLoS Genet 2008; 4:e1000111.
118. Thany SH, Gauthier M. Nicotine injected into the antennal lobes induces a rapid modulation of sucrose threshold and improves short-term memory in the honeybee Apis mellifera. Brain Res 2005; 1039:216-219.
119. Lozano VC, Armengaud C, Gauthier M. Memory impairment induced by cholinergic antagonists injected into the mushroom bodies of the honeybee. J Comp Physiol [A] 2001; 187:249-254.
120. Lozano VC, Bonnard E, Gauthier M et al. Mecamylamine-induced impairment of acquisition and retrieval of olfactory conditioning in the honeybee. Behav Brain Res 1996; 81:215-222.
121. Dacher M, Lagarrigue A, Gauthier M. Antennal tactile learning in the honeybee: effect of nicotinic antagonists on memory dynamics. Neuroscience 2005; 130:37-50.

122. Hogg RC, Raggenbass M, Bertrand D. Nicotinic acetylcholine receptors: from structure to brain function. Rev Physiol Biochem Pharmacol 2003; 147:1-46.
123. Rocher A, Marchand-Geneste N. Homology modelling of the Apis mellifera nicotinic acetylcholine receptor (nAChR) and docking of imidacloprid and fipronil insecticides and their metabolites. SAR QSAR Environ Res 2008; 19:245-261.
124. Altschul SF, Gish W, Miller W et al. Basic local alignment search tool. J Mol Biol 1990; 215:403-410.
125. Kondrashov FA, Koonin EV. Origin of alternative splicing by tandem exon duplication. Hum Mol Genet 2001; 10:2661-2669.

CHAPTER 4

Identification of Critical Elements Determining Toxins and Insecticide Affinity, Ligand Binding Domains and Channel Properties

Hélène Tricoire-Leignel and Steeve Hervé Thany*

Abstract

Insect nicotinic acetylcholine receptors have been objects of attention since the discovery of neonicotinoid insecticides. Mutagenesis studies have revealed that, although the detailed subunit composition of insect nicotinic acetylcholine receptors subtypes eludes us, the framework provided by mutagenesis analysis makes a picture of the subunits involved in the ligand binding and channel properties. In fact, many residues that line the channel or bind to the ligand seemed to be strongly conserved in particular in the N-terminal extracellular region and the second transmembrane domain which constitutes the ion-conducting pathway supporting the flux of ions as well as their discrimination. In fact, the positions are carried by loops B and C, respectively, which contain amino acids directly contributing to the acetylcholine binding site. Mutation of these residues accounts for insect resistance to neonicotinoid insecticides such as imidacloprid or a loss of specific binding. The discovery of the same mutation at homologous residues in different insect species or its conservation raises the intriguing question of whether a single mutation is essential to generate a resistance phenotype or whether some subunit confer insensitivity to ligand. Consequently, recent finding using information from *Torpedo marmorata* α1 subunit and soluble *Aplysia californica* and *Lymnae stagnalis* acetylcholine binding proteins from crystallization suggest that insect nAChR subunits had contributing amino acids in the agonist site structure which participate to affinity and pharmacological properties of these receptors. These new range of data greatly facilitate the understanding of toxin-nAChR interactions and the neonicotinoid binding and selectivity.

Introduction

Based on nAChR from electric organ of the marine ray *T. marmorata*, affinity labelling, mutagenesis and structural studies have provided compelling evidence to locate the agonist binding sites at the interface between subunits. Then, structural information available indicate that

*Corresponding Author: Steeve Hervé Thany—Laboratoire Récepteurs et Canaux Ioniques Membranaires (RCIM), UPRES EA 2647/USC INRA 2023, IFR 149 QUASAV, Université d'Angers, UFR de Sciences. 2 Bd. Lavoisier 49045 Angers, France.
Email: steeve.thany@univ-angers.fr

Insect Nicotinic Acetylcholine Receptors, edited by Steeve Hervé Thany.

vertebrate nAChRs is a 290 kDa, hetero-pentameric glycol-protein consisting of a cation-selective membrane-spanning pore. The nAChR subunit genes fall into two main classes: the α subunits which possess two adjacent cysteines essential for acetylcholine binding whereas the non-α referred do not. Consequently, in the line of studies on vertebrate neuronal nAChRs, several insect nAChR subunits have been cloned. Nowadays, more than sixty subunits sequences are available and alternative splicing and RNA editing mechanisms detected in α4,[1-3] α6[4,5] and α9 subunits[6] increase the number of potential subunits assembly in some insects. Each subunit provides three loops of amino acid residues which are involved in acetylcholine binding site: A, B and C for the principal part, associated with neighbouring subunit residues contributing to the D, E and F loops for the complementary part.[7-9] Interestingly, other residues localized outside of the loops that define this agonist binding site are also important for the ligand binding and ion selectivity. Most of these key residues were identified in the crystal structure of a soluble homopentameric acetylcholine binding protein (AChBP), which was solved from the glia cells of the freshwater snail *Lymnaea stagnalis*[10,11] and then from the saltwater mollusc *Aplysia californica*.[12] AChBP is a soluble protein homologue to the extracellular domain of nAChRs which shares virtually all the ligand-binding characteristics with the nicotinic receptors family and revealed a structure largely consistent with studies based on electron-microscopy, ligand chemical modifications and nAChR and/or ligand mutagenesis. As its sequence shares amino-acid identity with the amino-terminal part of vertebrate as insects nAChRs, this AChBP was considered as a reliable structure for nAChR homology modelling and docking simulations of nAChR ligands such as insecticides and toxins.[10,12-17]

Toxins—nAChRs Interaction

Identification of toxin specific pharmacophores was well described in vertebrate nAChRs for α-conotoxins as well as α-Bungarotoxine (α-Bgt) and α-Cobratoxin (α-CBX).[17-21] α-conotoxin ImI residues like aspartic acid in 5th location (D5), prolin in 6th location (P6) and arginin in 7th location (R7) exhibit Van der Waals interactions with the C loop-over-β-leaflet residues as tyrosin in 193rd location (Y193) and loop B tryptophan in 147th location (W147) of the α7 homomeric receptor (Fig. 1).[22] However, loop C Y193 needs to be associated to F185 and Y186 of α-Bgt and α-CBX, so that α7 receptor could bind both toxins.[17] Moreover, minus sequences variations between toxins from a same animal could modify the targeted nAChR subtype. Indeed, depending on the 10th amino-acid of α-conotoxins PnIA and PnIB, ligand-receptor binding occurs with loop C Y193 of α7 receptor (ex:PnIB), or with α3 subunit P180 of α3β2 receptor (ex:PnIA).[18] Nevertheless, N11 in α-PnIA combined with α3-subunit I186 seems to play a crucial role in the pharmacological characterization of α7 and α3β2 receptors.[18] In addition, α-PIA, a cone produced-toxin (*Conus purpuraceus),* could discriminate α3β2 and α6β2 heteromeric receptors.[23] Then, in binding sites originated in different subunits combination according to the receptor subtype, residues variations could predict the toxin type, which could potentially bind the receptor.

Insecticides—nAChRs Interaction: Residues Involved in Neonicotinoid Selectivity

In parallel with toxins, neonicotinoids do require specific residues to bind insect nAChRs. These residues could be different according to the molecular properties of each active compound and the subunits types engaged in binding sites. In order to enhance insecticide design, molecular modeling and virtual docking of insecticides have followed previous pharmacological studies based on sequence-modified subunits and/or chemical-modified active insecticide compound.

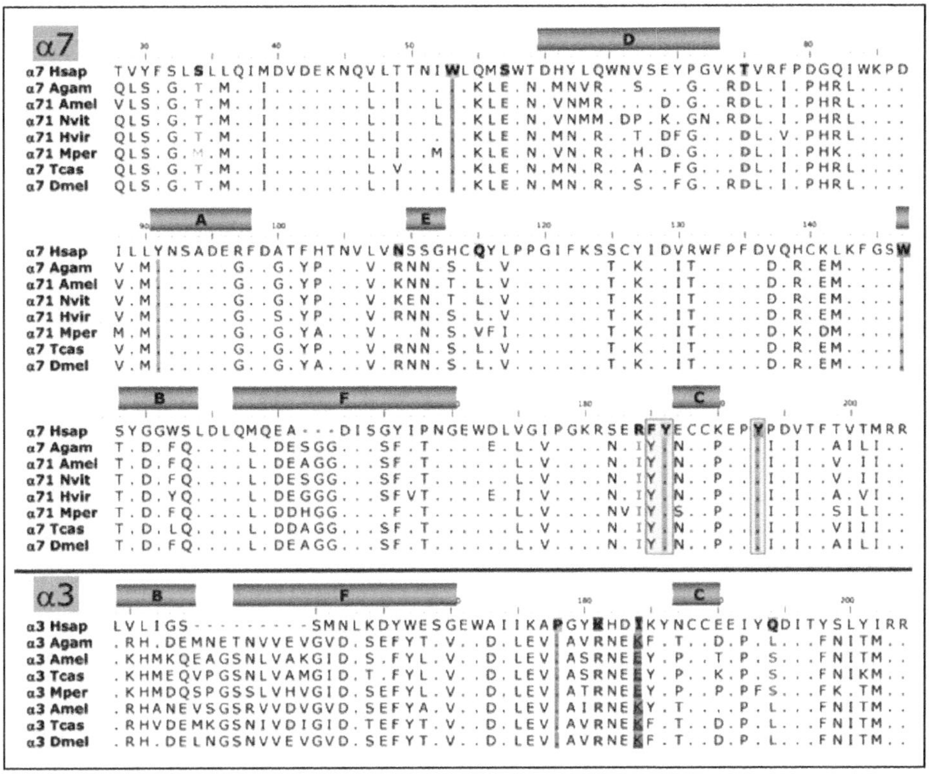

Figure 1. Extracellular domain partial amino-acid sequence alignment of nAChRs α3 and α7 subunits from human (Hsap: *Homo sapiens*) and seven insect species (Agam: *Anopheles gambiae*, Amel: *Apis mellifera*, Nvit: *Nasonia vitripennis*, Hvir: *Heliothis virescens*, Mper: *Myzus persicae*, Tcas: *Tribolium castaneum*, Dmel: *Drosophila melanogaster*). A, B, C, D, E and F boxes underline the loops position involved in acetylcholine binding site. Bold residues on human sequences correspond to various toxins binding sites (blue: α-ImI, orange: α-PnIB, green: α-PnIA, red: α-MII and in grey boxes: α-bgt et α-CBX). Some residues are well-conserved between species although others are substituted into unique or several amino-acids in insects. A color version of this image is available at www.landsbioscience.com/curie.

Neonicotinoid selectivity was attributed to insect-specific regions in α subunits (Fig. 2). Then, the loops A-C, the region between the loops B and C and between the loop B and N-terminus seem to be crucial for imidacloprid binding.[24-26] Acute experiments based on site-directed mutagenesis reveal that this insecticide binding is dependent from a proline in loop C in drosophila α2 subunit.[26] Indeed, when glutamate identified in loop C of chicken α4 subunit is substituted by the corresponding proline of drosophila α2 subunit, imidacloprid binding occurs in chimeric chicken α4β2 nAChR. Moreover, R77 and V79 in loop D of drosophila β subunits enhance imidacloprid binding when introduced in loop D of chicken β2 subunit instead of T77 and E79.[27] These results have been confirmed when the 10 residues-loop D from *Myzus persicae* or *Drosophila melanogaster* β1 subunit were introduced into the rat β2 subunit in nAChRs *Nilaparvata lugens* α1-*Rattus norvegicus* β2 chimeras, leading to increased imidacloprid affinity. Using the same approach, the authors underlined new key residues, i.e., Y131 or R131 and

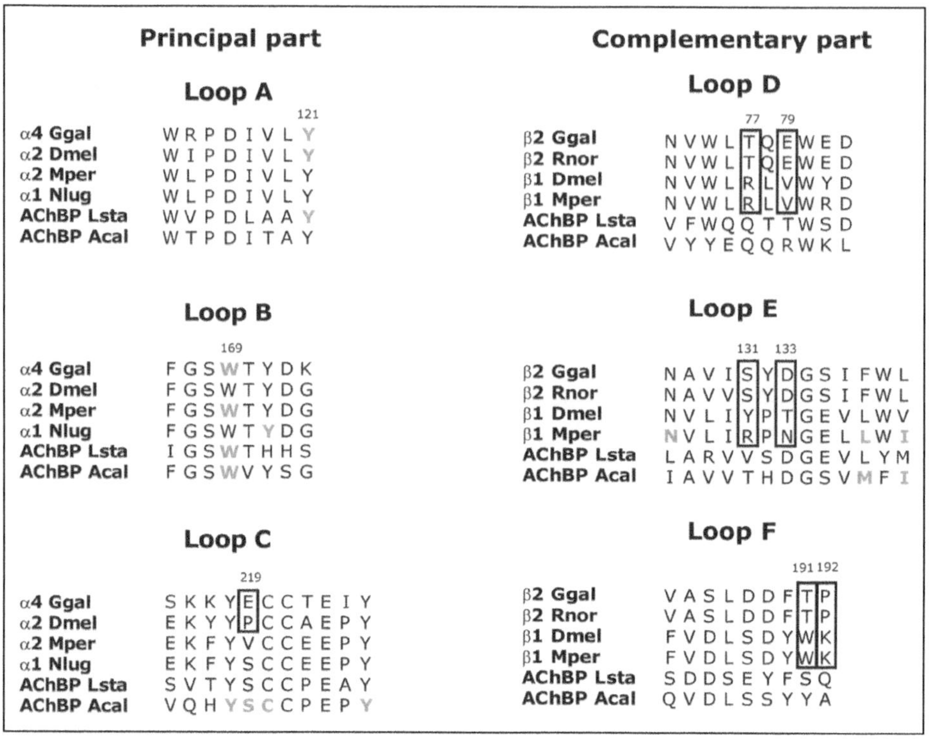

Figure 2. Amino-acid alignments of the loops A-F involved in ligand binding domain from vertebrates (Ggal: *Gallus gallus*; Rnor: *Rattus norvegicus*) and insects (Dmel: *Drosophila melanogaster*; Mper: *Myzus persicae*; Nlug: *Nilaparvata lugens*) in comparison with acetylcholine binding protein (AChBP) from *Lymnea stagnalis* (Lsta) and *Aplysia californica* (Acal), using ClustalW method. Boxes underline key residues in neonicotinoid selectivity; while orange bold residues account for neonicotinoid binding throughout the loops, according to mutagenesis experiments and insecticides docking on homologous models (see text for references). Numbers indicate above the alignments correspond to chicken subunits (α4 in principal part and β2 in complementary part). *Accession numbers*: α4 Ggal: NP990145, β2 Ggal: NP990144, β2 Rnor: NP062170, α2 Dmel: CAA36517, β1 Dmel: CAA27641, α2 Mper: CAA57477, β1 Mper: CAB87995, α1 Nlug: AAQ75737, AChBP Lsta: AAK64377, AChBP Acal: AAL37251. A color version of this image is available at www.landsbioscience.com/curie.

N133 instead of S131 and D133 in rat β2 subunit loop E as well as W191 and K192 instead of Y191 and P192 in loop F.[28] Moreover, combinatorial mutations affecting loops C and D of the chicken α4β2 nAChR enhance the agonist efficacy of imidacloprid, suggesting a synergistic effect of the two loops on the interactions with this insecticide when key insect nAChR residues replace the vertebrate equivalents.[29] This observation was assessed by computational modeling of the ligand binding domain of the wild-type and mutant α4β2 nAChRs with imidacloprid bound, using *Lymnae stagnalis* AChBP as template (Fig. 2).

Insecticides—nAChRs Interaction: Residues Involved in Neonicotinoid Binding

Because neonicotinoids were wide used to manage most destructive crop pests around the world, strong resistance in some of these species appeared. To design novel active molecules for crop protection, current neonicotinoid-nAChRs interactions are of considerable interest.[30]

Using brown planthopper populations differing in imidacloprid sensitivity, Liu and its colleagues detected a single mutation Y151S in the loop B of the α1 subunit conferring imidacloprid resistance. This mutation in loop B prevented imidacloprid binding on heteromeric receptors, assessing the key role of this residue in the insecticide binding.[31] Moreover, structure-activity relationship studies led to binding models of neonicotinoids implying H-donating and electron-rich sites in nAChRs, namely aromatic residue like tryptophan (W169) and positively charged residues like lysine and arginine residues (L78 and R79).[30,32] Indeed, recent cocrystallisation of the AChBP from *Lymnea stagnalis* (*Ls*-AChBP) with imidacloprid and clothianidin reveal that the Q55 in loop D of Ls-AChBP hydrogen bonded with the nitro group of imidacloprid and that the backbone carbonyl of W143 hydrogen bonds with the NH of the guanidine moiety of clothianidin.[33] In parallel, cocrystallisation of the AChBP from *Aplysia californica* (*Ac*-AChBP) with imidacloprid and thiacloprid suggests a neonicotinoid electronegative pharmacophore including W147 (loop B), Y188, S189, C190 and Y195 (loop C) from the principal face and Y55 (loop D), M116 and I118 (loop E) from the complementary face.[34] In order to dock insecticide molecules on interfacial agonist binding domain of an insect nAChR and not only a mollusk, a structural homology model of the peach-potato aphid α2β1 nAChR was built from the crystal structure of the *Ac*-AChBP, which is sensitive to neonicotinoid.[35] Docking simulations for several neonicotinoid analogues suggest that relevant amino-acids belonging to α-subunit loops B (W174) and C (Y224, C226 and C227) and the β subunit loops D (W79 and R81) and E (N131, L141 and I143) represent neonicotinoid binding pockets, which could allow discovery of novel insecticides.[35,36]

Although neonicotinoid compounds are selective for insects and not vertebrate, the side-effects of wide use in agricultural fields were pointed out namely on useful insects as pollinators like honeybees. Recent homology modeling of *Apis mellifera* α3β2, α4β2, α6β2, α1β1 ligand binding domain allowed docking studies for imidacloprid and associated metabolites.[37] The docked conformations of insecticides are different in each receptor subtype, underlining the crucial role of subunits types in the binding site. Indeed, when α4 is included in the binding site of imidacloprid, the number of hydrogen bonds reflecting the affinity is higher than α3 or α6 binding sites. However, 4OH-imidacloprid for example makes more hydrogen bonds with those last nAChR subtypes than in α4 binding site, indicating that some metabolites display higher number of hydrogen bonds than their parent compounds.

Amino-Acid Involved in Ionic Selectivity

Mutagenesis and substituted-cysteine-accessibility method (SCAM) studies have allowed identification of amino acids that contribute to both determining pharmacological diversity and charge selectivity filter.[38,39] This peculiar pattern of conservation led us to believe it could represent structures important for proper function. We therefore examined the conservation of these residues in the M1 and M2 transmembrane domains of insect nAChR subunits as these domains form the lumen of the channel.[8] The alignment of the M1 and M2 transmembrane domains reveals conservation of several residues within the M1-M2 loop and in the M2 transmembrane domain from vertebrate to insect, particularly, glycine, valine and acid glutamic at position 237, 251 and 258 respectively. The glycine and acid glutamic residues form the 'intermediate' rings of negatively charged amino acids as previously shown[40] and mutation of these residues to neutral amino acids led to functional receptors which are selective for cation,[38] demonstrating that this change could modify the ion channel properties. The mutation of this single valine V251 by threonine in the vertebrate neuronal α7 nAChR subunit (mutation V251T) accounts for the changes in apparent affinity for ACh, sensitivity to DHβE, response desensitization and current rectification.[38] Note that this valine was replaced in methionine in *Apis mellifera* Apisα2, *Drosophila melanogaster* Dα2 and *Myzus persicae* Mpα1. Moreover, the leucine residue at position 247 has been suggested to be conserved in all Ligand Gated Ion Channel Cys-loop family[41] and plays a pivotal role in the properties of the receptor.[42] Mutation of this leucine in the M2 transmembrane domain alters the desensitization, Ca^{2+} permeability and pharmacological profile of α7 nAChRs.[43-45] Interestingly this leucine was mutated in valine in *M. persicae* Mpα4 and in isoleucine in *D. melanogaster* Dα4, *Locusta migratoria* Locα1 and *M. persicae* Mpα5. Although, no direct evidence has been done in insect due in part to difficulty to express functional insect nAChR subunits in heterologous system, we could suggest that the mutation of these residues in insect could account for specific pharmacology. The analysis of naturally occurring mutations in insect subunits is an important source of information on structure-function relationship and it could explain why imidacloprid bind *Periplaneta americana* nAChR2 only when the channel lumen is open by agonist action.[46]

Conclusion

Structure-activity relationship studies from last decade were the keys in understanding interaction between ligands as toxins or insecticides and specific nAChRs, as well as key residues from the M2 transmembrane segment which delimits the pore region of the cation-selective channel. Indeed, methodical amino-acid substitutions in a specific nAChRs subunit from homomeric receptor reveal crucial amino-acids for ligand-receptor interaction and ionic selectivity. Then, combined data from directed mutagenesis targeting the nAChR allow the design of ligand-specific pocket, named as pharmacophore as well as the charge selectivity filter. Moreover, the discovery and crystallization of soluble AChBP from *Aplysia californica* and *Lymnae stagnalis* greatly facilitate the understanding of toxin-nAChR interactions and the neonicotinoid binding and selectivity.

The molecular basis for the disparate subunit composition of insect nAChR subtypes is not known but it is usually assumed that the ACh binds at the interfaces of the α subunits with neighbouring α or nonα (β) subunits. The data exposed above highlight the complexity of ligand binding pocket, according to the subunit arrangement forming the binding site and the agonist molecule, i.e., toxins or insecticides. However, structural features as extracellular loops remain involve in binding site, but key residues belonging to N-terminal loops vary according to agonist molecule chemical properties as well as engaged subunits in binding sites. Concerning insecticide selectivity, *Ls*-AChBP is less sensitive to neonicotinoid than *Ac*-AChBP is, suggesting pharmacological profiles reminiscent of vertebrate and insect nAChRs, respectively. As these two AChBP subtypes (*Ac*-AChBP and *Ls*-AChBP) have distinct pharmacology toward agonists when expressed and purified in vitro, comparison of pharmacophores would help us understanding the chemical features conferring insecticide selectivity.

The homology modeling using conformational information of both AChBP and marine ray α1 subunit give us new perspectives namely in discovering insect nAChRs. Thus, specific and exclusive insect binding sites can be created as insect subunits alone are involved in virtual arrangement. This was not able in most of hybrid recombinant nAChRs including vertebrate subunits, which are expressed in heterologous system as *Xenopus* oocytes or lineage cells. Nevertheless, comparison of data from both techniques will give us a better knowledge of insect nAChRs pharmacological properties and diversity.

References

1. Jones AK, Sattelle DB. The cys-loop ligand-gated ion channel superfamily of the honeybee, Apis mellifera. Invert Neurosci 2006; 6(3):123-132.
2. Jones AK, Sattelle DB. The cys-loop ligand-gated ion channel gene superfamily of the red flour beetle, Tribolium castaneum. BMC Genomics 2007; 8:327.
3. Lansdell SJ, Millar NS. Cloning and heterologous expression of Dalpha4, a drosophila neuronal nicotinic acetylcholine receptor subunit: Identification of an alternative exon influencing the efficiency of subunit assembly. Neuropharmacology 2000; 39(13):2604-2614.

4. Grauso M, Reenan RA, Culetto E et al. Novel putative nicotinic acetylcholine receptor subunit genes, Dalpha5, Dalpha6 and Dalpha7, in Drosophila melanogaster identify a new and highly conserved target of adenosine deaminase acting on RNA-mediated A-to-I premrna editing. Genetics 2002; 160(4):1519-1533.
5. Sattelle DB, Jones AK, Sattelle BM et al. Edit, cut and paste in the nicotinic acetylcholine receptor gene family of Drosophila melanogaster. Bioessays 2005; 27(4):366-376.
6. Jones AK, Buckingham SD, Brown LA et al. Alternative splicing of the Anopheles gambiae nicotinic acetylcholine receptor, Agam alphabeta9, generates both α and β subunits. Invert Neurosci 2009; 9(2):77-84.
7. Arias HR. Localization of agonist and competitive antagonist binding sites on nicotinic acetylcholine receptors. Neurochem Int 2000; 36(7):595-645.
8. Corringer PJ, Le Novere N, Changeux JP. Nicotinic receptors at the amino acid level. Annu Rev Pharmacol Toxicol 2000; 40:431-458.
9. Unwin N. Projection structure of the nicotinic acetylcholine receptor: Distinct conformations of the α subunits. J Mol Biol 1996; 257(3):586-596.
10. Brejc K, van Dijk WJ, Klaassen RV et al. Crystal structure of an ACh-Binding Protein reveals the ligand-binding domain of nicotinic receptors. Nature 2001; 411(6835):269-276.
11. Smit AB, Syed NI, Schaap D et al. A glia-derived acetylcholine-binding protein that modulates synaptic transmission. Nature 2001; 411(6835):261-268.
12. Hansen SB, Talley TT, Radic Z et al. Structural and ligand recognition characteristics of an AcetylCholine-Binding Protein from Aplysia californica. J Biol Chem 2004; 279(23):24197-24202.
13. Celie PH, van Rossum-Fikkert SE, van Dijk WJ et al. Nicotine and carbamylcholine binding to nicotinic acetylcholine receptors as studied in AChBP crystal structures. Neuron 2004; 41(6):907-914.
14. Hibbs RE, Talley TT, Taylor P. Acrylodan-conjugated cysteine side chains reveal conformational state and ligand site locations of the acetylcholine-binding protein. J Biol Chem 2004; 279(27):28483-28491.
15. Sine SM. The nicotinic receptor ligand binding domain. J Neurobiol 2002; 53(4):431-446.
16. Tomizawa M, Casida JE. Molecular recognition of neonicotinoid insecticides: The determinants of life or death. Acc Chem Res 2009; 42(2):260-269.
17. Tsetlin VI, Hucho F. Snake and snail toxins acting on nicotinic acetylcholine receptors: Fundamental aspects and medical applications. FEBS Lett 2004; 557(1-3):9-13.
18. Dutertre S, Lewis RJ. Computational approaches to understand alpha-conotoxin interactions at neuronal nicotinic receptors. Eur J Biochem 2004; 271(12):2327-2334.
19. Ulens C, Hogg RC, Celie PH et al. Structural determinants of selective α-conotoxin binding to a nicotinic acetylcholine receptor homolog AChBP. Proc Natl Acad Sci USA 2006; 103(10):3615-3620.
20. Dutertre S, Ulens C, Buttner R et al. AChBP-targeted α-conotoxin correlates distinct binding orientations with nAChR subtype selectivity. EMBO J 2007; 26(16):3858-3867.
21. Bourne Y, Talley TT, Hansen SB et al. Crystal structure of a Cbtx-AChBP complex reveals essential interactions between snake alpha-neurotoxins and nicotinic receptors. EMBO J 2005; 24(8):1512-1522.
22. Terlau H, Olivera BM. Conus venoms: A rich source of novel ion channel-targeted peptides. Physiol Rev 2004; 84(1):41-68.
23. Janes RW. Alpha-conotoxins as selective probes for nicotinic acetylcholine receptor subclasses. Curr Opin Pharmacol 2005; 5(3):280-292.
24. Matsuda K, Shimomura M, Kondo Y et al. Role of loop D of the α7 nicotinic acetylcholine receptor in its interaction with the insecticide imidacloprid and related neonicotinoids. Br J Pharmacol 2000; 130(5):981-986.
25. Shimomura M, Satoh H, Yokota M et al. Insect-vertebrate chimeric nicotinic acetylcholine receptors identify a region, loop B to the N-terminus of the drosophila Dα2 subunit, which contributes to neonicotinoid sensitivity. Neurosci Lett 2005; 385(2):168-172.
26. Shimomura M, Yokota M, Matsuda K et al. Roles of loop C and the loop B-C interval of the nicotinic receptor alpha subunit in its selective interactions with imidacloprid in insects. Neurosci Lett 2004; 363(3):195-198.
27. Shimomura M, Yokota M, Ihara M et al. Role in the selectivity of neonicotinoids of insect-specific basic residues in loop D of the nicotinic acetylcholine receptor agonist binding site. Mol Pharmacol 2006; 70(4):1255-1263.
28. Yao X, Song F, Chen F et al. Amino acids within loops D, E and F of insect nicotinic acetylcholine receptor beta subunits influence neonicotinoid selectivity. Insect Biochem Mol Biol 2008; 38(9):834-840.

29. Toshima K, Kanaoka S, Yamada A et al. Combined roles of loops C and D in the interactions of a neonicotinoid insecticide imidacloprid with the α4β2 nicotinic acetylcholine receptor. Neuropharmacology 2009; 56(1):264-272.
30. Jeschke P, Nauen R. Neonicotinoids-from zero to hero in insecticide chemistry. Pest Manag Sci 2008; 64(11):1084-1098.
31. Liu Z, Williamson MS, Lansdell SJ et al. A nicotinic acetylcholine receptor mutation conferring target-site resistance to imidacloprid in Nilaparvata lugens (brown planthopper). Proc Natl Acad Sci USA 2005; 102(24):8420-8425.
32. Wang Y, Cheng J, Qian X et al. Actions between neonicotinoids and key residues of insect nAChR based on an ab initio quantum chemistry study: Hydrogen bonding and cooperative pi-pi interaction. Bioorg Med Chem 2007; 15(7):2624-2630.
33. Ihara M, Okajima T, Yamashita A et al. Crystal structures of Lymnaea stagnalis AChBP in complex with neonicotinoid insecticides imidacloprid and clothianidin. Invert Neurosci 2008; 8(2):71-81.
34. Talley TT, Harel M, Hibbs RE et al. Atomic interactions of neonicotinoid agonists with AChBP: Molecular recognition of the distinctive electronegative pharmacophore. Proc Natl Acad Sci USA 2008; 105(21):7606-7611.
35. Ohno I, Tomizawa M, Durkin KA et al. Molecular features of neonicotinoid pharmacophore variants interacting with the insect nicotinic receptor. Chem Res Toxicol 2009.
36. Ohno I, Tomizawa M, Durkin KA et al. Bis-neonicotinoid insecticides: Observed and predicted binding interactions with the nicotinic receptor. Bioorg Med Chem Lett 2009; 19(13):3449-3452.
37. Rocher A, Marchand-Geneste N. Homology modelling of the Apis mellifera nicotinic acetylcholine receptor (nAChR) and docking of imidacloprid and fipronil insecticides and their metabolites. SAR QSAR Environ Res 2008; 19(3-4):245-261.
38. Galzi JL, Devillers-Thiery A, Hussy N et al. Mutations in the channel domain of a neuronal nicotinic receptor convert ion selectivity from cationic to anionic. Nature 1992; 359(6395):500-505.
39. Corringer PJ, Bertrand S, Bohler S et al. Critical elements determining diversity in agonist binding and desensitization of neuronal nicotinic acetylcholine receptors. J Neurosci 1998; 18(2):648-657.
40. Imoto K, Busch C, Sakmann B et al. Rings of negatively charged amino acids determine the acetylcholine receptor channel conductance. Nature 1988; 335(6191):645-648.
41. Jensen AA, Frolund B, Liljefors T et al. Neuronal nicotinic acetylcholine receptors: Structural revelations, target identifications and therapeutic inspirations. J Med Chem 2005; 48(15):4705-4745.
42. Buckingham SD, Adcock C, Sansom MS et al. Functional characterization of a mutated chicken alpha7 nicotinic acetylcholine receptor subunit with a leucine residue inserted in transmembrane domain 2. Br J Pharmacol 1998; 124(4):747-755.
43. Bertrand D, Devillers-Thiery A, Revah F et al. Unconventional pharmacology of a neuronal nicotinic receptor mutated in the channel domain. Proc Natl Acad Sci USA 1992; 89(4):1261-1265.
44. Palma E, Mileo AM, Eusebi F et al. Threonine-for-leucine mutation within domain M2 of the neuronal alpha(7) nicotinic receptor converts 5-hydroxytryptamine from antagonist to agonist. Proc Natl Acad Sci USA 1996; 93(20):11231-11235.
45. Palma E, Fucile S, Barabino B et al. Strychnine activates neuronal alpha7 nicotinic receptors after mutations in the leucine ring and transmitter binding site domains. Proc Natl Acad Sci USA 1999; 96(23):13421-13426.
46. Courjaret R, Lapied B. Complex intracellular messenger pathways regulate one type of neuronal alpha-bungarotoxin-resistant nicotinic acetylcholine receptors expressed in insect neurosecretory cells (Dorsal Unpaired Median neurons). Mol Pharmacol 2001; 60(1):80-91.

Chapter 5

Electrophysiological Studies and Pharmacological Properties of Insect Native Nicotinic Acetylcholine Receptors

Steeve Hervé Thany*

Abstract

The existence of several nicotinic acetylcholine receptor genes in insects suggests that many nicotinic receptor subtypes are present, but the identification and characterization of these subtypes in native neurons has been limited. Their pharmacological properties came from electrophysiological studies in which variations in the sensitivity of insect neurons were correlated with time course, current amplitudes, desensitization rates occurring in varying proportions in different cells. Thus pressure application of agonists on cultured cells induced inward currents showing that acetylcholine and nicotine were partial agonists of some cells with a lower efficacy while they were full agonists in other neurons. The variation in kinetics appeared to be due to differential expression of distinct nicotinic receptor subtypes as corroborated by the blocking activity induced by antagonists. In fact, the alpha-bungarotoxin-sensitive nicotinic receptor subtypes described as homomeric could be also heteromeric receptors. Interestingly, some receptors mediating nicotinic responses have been termed 'mixed' receptors because they were blocked by a range of nicotinic and muscarinic antagonists.

Following electrophysiological studies, it has been also demonstrated that insect nicotinic receptors were modulated by Ca^{2+} pathways. Ca^{2+} permeability through insect nicotinic receptors, voltage-gated Ca^{2+} channels or released from intracellular stores represents an important indication of insect native nicotinic acetylcholine receptor modulation. The Ca^{2+} flow may trigger a variety of cytosolic Ca^{2+} pathways underlying many cellular processes such Calmodulin kinase, PKA and PKC. Most of the studies suggested that the effect of phosphorylation mechanism was dependent on the receptor subtype.

Introduction

The cell body membrane of insect presents a preparation in which the characteristics of insect nicotinic neurons can be studied under voltage- and current-clamp techniques. A population of alpha-bungarotoxin (α-Bgt)-sensitive and –insensitive nicotinic acetylcholine receptor (nAChR)

*Steeve Hervé Thany—Laboratoire Récepteurs et Canaux Ioniques Membranaires (RCIM), UPRES EA 2647/USC INRA 2023, IFR 149 QUASAV, Université d'Angers, UFR de Sciences. 2 Bd. Lavoisier 49045 Angers, France. Email: steeve.thany@univangers.fr

Insect Nicotinic Acetylcholine Receptors, edited by Steeve Hervé Thany.

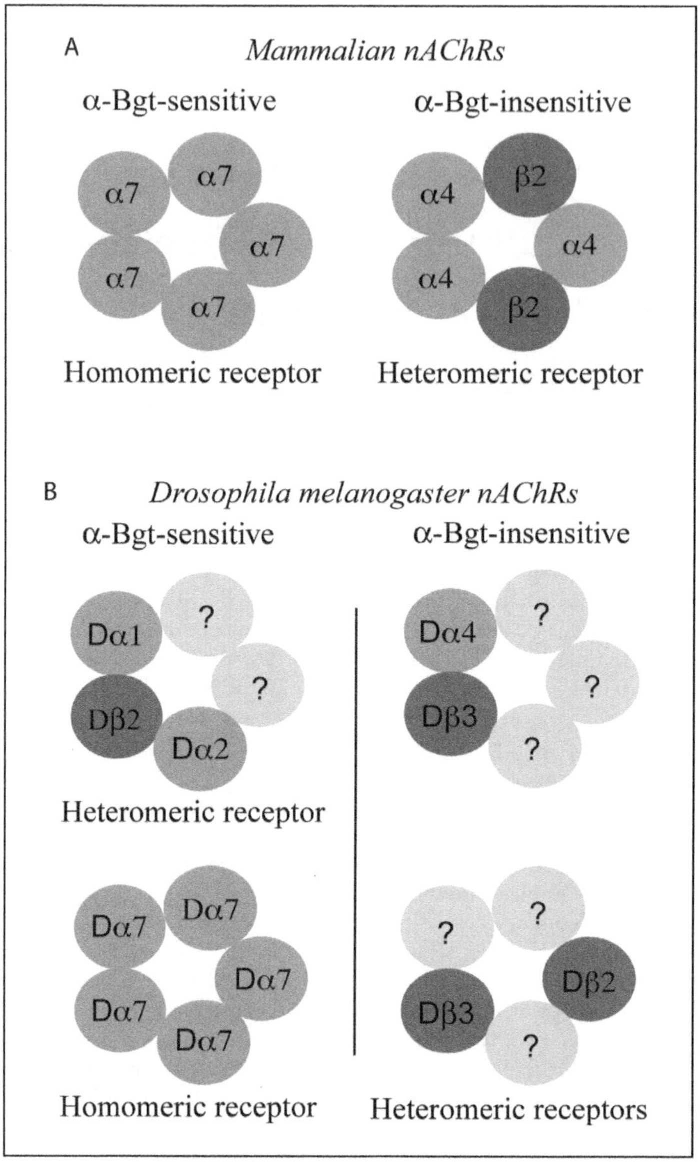

Figure 1. Insect nAChR subtypes. As vertebrates there are a pentameric receptors composed of five subunits. α-Bgt-sensitive nAChR subtypes can be heteromeric or homomeric receptors.

subtypes have been identified.[1] As vertebrate, it has been suggested that α-Bgt-sensitive nAChR subtypes were homomeric receptors while α-Bgt-insensitive subtypes were heteromeric receptors (Fig. 1). In fact, nicotine when applied to this neuron resulted in the generation of an inward current which was blocked by α-Bgt and in some case remained insensitive to this toxin. Based on both these two subtypes and several nicotinic antagonists, native neuronal nAChR subtypes have been described in several insect species such as honeybees,[2-5] cockroaches,[6-10] crickets,[11] drosophila[12-14] and locusts[15] (Table 1). In all cases, the mean current amplitudes suggested that

Table 1. Pharmacological profiles on native cells

Species	Identified Cell Types	Pharmacological Profiles		References
		Agonist	Antagonist	
Periplaneta americana	-DUM neurons -Fast coxal depressor motoneuron -Thoracic ganglia neurons	ACh, Nicotine, Imidacloprid (inhibit the desensitized component), Clothianidin, Epibatidine.	d-TC (block nAChR1), Meca (Block nAChR2), α-Conotoxine, α-Bgt, DHβE, MLA (inhibit the nondesensi tized component)	7, 8, 52, 25, 10
Apis mellifera	-Kenyon cells -Antennal lobes neurons	ACh, CCh, Imi (partial agonist), Nic, Epi, Cytisine, Olefine, 5-OH-Imi	α-Bgt, DHβE. Meca, d-TC	3, 4, 26, 27
Drosophila melanogaster	-Kenyon cells -Clock neurons -Ventral nerve cord neurons	ACh, Nic	α-Bgt, d-TC, Meca	12, 13, 14
Manduca sexta	-Abdominal ganglia neurons	ACh, Nic	Meca	71, 72, 24
Heliothis virescens	Ventral nerve cord neurons	N-des, Clothianidin, Imi	nd	73
Locusta migratoria	Thoracic ganglia neurons	ACh, Physostigmine (partial competitive agonist)	d-TC, Bicuculline, hydrastine, gabazine (partial inhibitor)	55

ACh : acetylcholine; DUM : dorsal unpaired median; Nic : nicotine; d-TC : d-tubocurarine; Meca : mecamylamine; CCh : carbamylcholine; Imi : imidacloprid; Epi : epibatidine; N-des : N-desmethyl thiamethoxam; DHβE : dihydro-β-erythroidine; MLA : methyllicaconitine; nd : not determined.

nicotinic agonists could act as partial agonist, full or, in some case remained a poor agonist of insect nAChRs expressed in the isolated cell. These pharmacological profiles were also highlight by behavioral studies using nicotine and nicotinic antagonists[16,17] or nicotinic acetylcholine receptor mutant[18] which have established that nAChR subtypes are differently involved in behavioral processes.[18,19] Finally, because several nAChR subunits have been identified in the insects central nervous system (CNS) from larval to adult stage,[1,20,21] the variable pharmacological properties (conductance, ion selectivity, affinity to the ligand) was due to combinatorial assembly of nAChR subunits which produce a wide structural diversity of receptors oligomer.

At functional level, insect α7-like subunits (e.g., Dα5, Dα6 and Dα7 subunits of *Drosophila*) which were potential candidates to form α-Bgt-sensitive receptors, could form heteromeric receptor in native cells. In addition, Dα1, Dα2, Dα3, Dβ1 and Dβ2 subunits can be copurified by α-Bgt affinity chromatography suggesting that (1) nAChRs composed to these subunits can bind α-Bgt and (2) they can form heteromeric α-Bgt-sensitive receptors.[22,23] Thus, insect native α-Bgt-sensitive receptors could contribute to either heteromeric and homomeric receptors (Fig. 1). Functionally, nAChRs are ligand-gated cationic channels with the capacity to elicit local changes in cytoplasmic calcium (Ca^{2+}) levels but the cellular mechanisms involved are not fully understood. Various Ca^{2+}-dependent proteins may be involved in the mechanisms regulating insect nAChR function such as calcium/calmodulin-dependent protein kinase (CaMKII), cyclic AMP-dependent protein kinase (PKA), protein kinase C (PKC) and endogenous protein tyrosine kinase.[8,9,24]

In this part, a combination of data on the native cells with electrophysiological findings from different species may provide some valuable insights in pharmacological properties and intracellular mechanisms regulating insect nAChRs.

Pharmacological Profiles of Native Nicotinic Receptors Associated to Specific Neurons

The more comprehensive understanding on the insect native nAChR subtypes and also their complexity came from eletrophysiological studies using different nAChR ligands. In vitro patch-clamp studies on the somata of different insect neurons showed that, many insect nAChRs share several properties with vertebrate neuronal nAChRs, such as their pharmacology, cation selectivity and current inward rectification. They are sensitive to nicotinic agonists nicotine, cytisine, epibatidine and to the nicotinic antagonists α-bungarotoxin (α-Bgt), methyllycaconitine (MLA), mecamylamine (MEC) and dihydroxy-β-erythroidine (DβHE). Generally, two distinct insect nAChR subtypes have been described, sensitive to the snake toxin α-Bgt and insensitive to this toxin.[2] This suggested that the first subclass was closed to the mammalian homomeric α7 receptors and the second was heteromeric receptors (Fig. 1). Nevertheless, recent studies gave opportunities to more understand the nAChR subtypes present in distinct insect neurons and their complexity. The first studies came from the cockroach *Periplaneta americana* thoracic ganglion neurons. Two α-Bgt-sensitive nAChR subtypes have been identified as desensitizing (nAChD) selectively inhibitable with imidaclopird and nondesensitizing (nAChN), selectively inhibitable with MLA.[25] In fact, MLA and α-Bgt preferentially blocked the nAChN current, while DβHE showed some selectivity for the nAChD current.[25] These two α-Bgt-sensitive receptors were distinct from the two α-Bgt-insensitive nAChRs isolated from cockroach dorsal unpaired median (DUM) neurons.[7-9] Under α-Bgt treatment, two distinct nAChR subtypes named nAChR1 and nAChR2 have been characterized.[7] They differed in their ionic permeability and pharmacology. For example, the antagonist d-tubocurarine (d-TC) can block nAChR1 but not nAChR2 which is blocked by MEC and the α7 nAChR-specific inhibitor, α-conotoxin.[7] These results highlight the finding that insect nAChR subtypes have somewhat different pharmacological profiles from mammalian nAChR subtypes and that cockroach expressed several nAChR subtypes which were differently expressed on thoracic and abdominal ganglia. This discrepancy was also identify in the honeybee *Apis mellifera* in which Kenyon cells neurons revealed different pharmacological properties to nicotinic agonists and antagonists.[2-5,26] Using Kenyon cells from honeybee pupae, Goldberg et al, demonstrated that approximately 80% of the ACh-induced current was irreversibly blocked by α-Bgt while atropine did not block these currents.[2] They concluded that this pharmacological profile defined a nAChR in which a large current may flow through α-Bgt-sensitive nAChRs and a smaller current through α-Bgt-insensitive nAChRs. Using the same cells, Wüstenberg and Grünewald demonstrated that carbamylcholine was a full agonist of the nicotinic receptor with a different dose-response relationships as compared to those of ACh-induced currents, while nicotine and epibatidine were only partial agonists.[4] Each antagonists tested, DβHE, MLA and MEC completely blocked the ACh-induced currents with different potencies.[4] These results indicated the expression of distinct receptor subtypes with differential sensitivity to nicotinic agonists and antagonists. Moreover, studies of nAChRs on cultured antennal lobe neurons from adult honeybee brains revealed that 90% of the cells responded to ACh application and 10% of the cells had low affinity to Ach.[5] Nicotine and imidacloprid elicited 45% and 43% of the maximum ACh-induced currents, respectively. They suggested that nAChRs from adult antennal lobes cells are different to nAChRs on pupal Kenyon cells by their pharmacological profile and ionic permeability. In fact, on adult cells the antagonist action of α-Bgt was fully reversible whereas on pupal cells it was only partially reversible. Nicotine and imidacloprid induced slightly smaller and less variable currents on pupal antennal lobes cells.[5,27] Their results were exemplified by the finding that depending on the antennal lobe neurons, imidacloprid either acted as a full agonist or a partial agonist[26] or as a partial agonist of pupal Kenyon cells.[3] In addition, *in situ* hybridation studies demonstrated that nAChR subunits were differently expressed between pupal and adult stages.[20] All these results confirmed that there was a specific expression of insect nAChR subtypes in some tissue.

Contribution of 'Mixed' Nicotinic/Muscarinic Receptor to the Complexity of Native Nicotinic Receptors

In insect, a mixed nicotinic/muscarinic acetylcholine receptor has been identified on the cell bodies of the fast coxal depressor motoneurons (Df) and dorsal unpaired median (DUM) neurons of the cockroach *Periplaneta americana*.[6,28] The cell body of the Df, third thoracic ganglion, displays two distinct types of electrical activity: (1) in response to long-duration depolarizing pulses or synaptic stimulation, it can generate plateau potentials and (2) it can generate calcium-dependent action potentials.[29-33] ACh when applied to this neuron voltage-clamped, results in the generation of an inward current which is blocked by α-Bgt. This α-Bgt-sensitive receptor was also blocked by a range of muscarinic antagonists such as pirenzepine, atropine and scopolamine. The receptors mediating this response have been termed 'mixed' cholinergic receptor.[28] Muscarinic agonists were found to mimic the α-Bgt-resistant component of the ACh response, for example the nonselective muscarinic agonist oxotremorine when applied to the preparation induced a change in the current/voltage relationship qualitatively similar to that caused by ACh in the presence of α-Bgt.[28] Cockroach DUM neurons, like motoneuron Df, possesses 'mixed' receptors which were sensitive to α-Bgt. In DUM neurons, ACh and nicotine induced depolarizing responses in which the slow component was sensitive to α-Bgt, d-tubocurarine, pirenzepine and gallamine, whereas the fast component was insensitive to these nicotinic and muscarinic antagonists.[6] These two distinct functional receptors, a sensitive nicotinic and a 'mixed' receptors were components of a nicotine-induced biphasic response.[6] A physiological role for these cholinergic receptors showing a 'mixed' activity could be to regulate transmitter release because there is evidence that receptors sensitive to muscarinic agonists may be involved in driving rhythmic motor activity, to drive burst of activity and rhythmic depolarizations in motoneurons. It seems that neurons from other insects possess a population of nicotinic receptors which are sensitive to both nicotinic and muscarinic agonists. In fact, a similar 'mixed' receptor has been identified in neurons from the thoracic ganglion of the locust *Locusta migratoria*.[34] It was not clear that these 'mixed' receptors are composed to the same or different subunits. Interestingly, a vertebrate homomeric α9 receptor showing a nicotinic and muscarinic pharmacology has been identified in the cochlear outer hair cells.[35,36] The α9 subunit can form an heteromeric receptor with the α10 subunit which displays faster and more extensive agonist-mediated desensitization and a biphasic response to changes in extracellular Ca^{2+} ions.[37] Both α9 and α10 genes exhibited a restricted expression pattern in the cochlear and vestibular hair cells suggesting that they participate in the efferent modulation of the cochlear amplifier and the control of the dynamic range of hearing.[35,38,39] It was noted that the pharmacological profiles of homomeric α9 and heteromeric α9α10 nAChRs are essentially indistinguishable.[37]

Ca^{2+} and Ca^{2+} Pathways as Intracellular Regulators of Insect Neuronal Nicotinic Receptors

Several studies have shown that the function of neuronal nAChRs is modulated by a variety of compounds including Ca^{2+} ions which act as nAChR allosteric modulator.[40,41] nAChR stimulation induced Ca^{2+} influx through nAChRs[42,43] or indirectly by the activation of voltage-operated Ca^{2+} channels.[44] Studies from insect nAChRs described an increase of intracellular Ca^{2+} concentration after ACh application which was associated to nAChR subtype. For example, down-regulation of MARA1 subunit mRNA significantly affected Ca^{2+} influx suggesting that nAChR subtypes composed to this subunit are involved in this mechanism.[24,45] Similarly, in the cockroach *Periplaneta americana* DUM neurons, application of nicotine induced an increase in intracellular Ca^{2+}-free concentration.[46] Because, Ca^{2+} responses to bath-applied nicotine were completely blocked by α-Bgt and partially by pirenzepine, this transient increase of Ca^{2+} was associated to 'mixed' receptor.[46] DUM neurons display a pacemaker activity that involves a variety of voltage-gated Ca^{2+} currents.[47,48] Nicotine in these cells modified the inactivation properties of the maintained low-voltage-activated Ca^{2+} currents and this effect was blocked when DUM neurons were pretreated with α-Bgt.[46] Moreover, it has been found that extracellular calcium influx through plasma membrane calcium channels modulated α-Bgt-insensitive nAChR2-mediated nicotine responses.[9] These results dedmonstrated that intracellular Ca^{2+} changes in DUM neurons could be associated to specific nAChR subtypes.

Ca^{2+} is a ubiquitous intracellular signal responsible for controlling numerous cellular processes which interacts with many other signalling pathways.[49] Variation in intracellular calcium are detected by multiple calcium-sensing proteins such as CaMKII, PKA and PKC.[44,49] Investigations performed on cockroach DUM neurons demonstrated that nAChR1 is the only one nicotinic receptor modulated by intracellular messenger such as cAMP, PKA or CaMKII.[7] Its function has been seen to be up- and down-regulated by two PKC that differ in their pharmacological properties and intracellular calcium sensitivity. The PKC1 which is activated by the phorbol-12-myristate-13-acetate, insensitive to rottlerin, is dependent on intracellular calcium and the PKC2 activated by the diacylglycerol analogue DiC8 and inhibited by rottlerin is calcium-independent.[8] These results were consistent with vertebrates studies showing the existence of several PKCs.[50,51] The PKC family comprised at least 11 isozymes divided into 3 subfamilies, conventional, novel and atypical which have distinct activation mechanisms and requirements.[52] The activation of conventional PKC requires coincident elevations of both intracellular Ca^{2+} concentration and diacylglycerol and proceeds via a series of sequential reactions.[50,51] It seems that PKCs activate via a direct or indirect pathways nAChR1 and intracellular Ca^{2+} modulates nAChR2.

In *Drosophila*, two distinct strains, *Dunce* which have mutations in the gene encoding the Type IV cAMP phosphodiesterase (PDE) and *DCO* which have mutations in the gene encoding the major catalytic subunit of PKA, were used to study the involvement of the cAMP/PKA cascade in the regulation of nicotine-induced effects.[53] Single exposure to nicotine dose-dependently inhibited the startle-induced climbing response. This effect was stronger in *Dunce* and wild-type, which have defective PDE, whereas it was weaker in *DCO* and wild-type which have defective PKA.[53] Analysis of the second messenger system such as calmodulin or protein kinases which up and down-regulate insect nAChRs and changes in intracellular Ca^{2+}, known to occur in insect neurons following nAChR activation demonstrates that a differential modulation of the excitability of the cell bodies and nerve terminals can occur depending on the nAChR subtypes expressed in the insect neurons.

Other Modulators of Insect Native Nicotinic Acetylcholine Receptors

Dopamine, octopamine and 5-hydroxytryptamine (5-HT) are widespread in the central nervous system[54-56] and have been shown to modulate nAChR responses.[57,58] They reversibly suppressed ACh responses indicating that they did not act by accelerating ACh degradation by acetylcholinesterases but their action was receptor-mediated because it can be blocked by pharmacological antagonists of monoamine receptors.[57] 5-HT was the most modulator which exerts its full effect upon ACh responses. This modulatory effect involved phosphorylation mechanisms. In fact, protein kinase inhibitors significantly attenuate modulation whereas suppression of ACh responses by 5-HT is blocked by specific competitive inhibitors of PKA and PKG.[58] Interestingly, the magnitude of the modulatory effect of 5-HT was significantly reduced by intracellular guanosine-5'-O-(2-thiodiphosphate) (GDP-β-S). GDP-β-S is a GDP analogue which prevents G-protein activation by competing with endogenous GTP for the guanine nucleotide binding site. The results indicated that 5-HT activate cGMP pathways via G-proteins confirming that analogues of both cAMP and cGMP can mimic the effect of 5-HT on nACh currents.[57,59]

An additional dimension of the insect nAChRs modulation was that it could be mediated following GABAergic responses. Synaptic inhibition is a major requirement for proper brain function and GABA is the major inhibitory neurotransmitter in the central nervous system of vertebrates and insects. Bicuculline, a specific $GABA_A$ receptor antagonist has been found to block neuronal nAChRs.[60-63] This effect was shown on both neurite and cockroach cell body receptors of giant interneuron and motor neuron D_f. From vertebrate studies, it seemed that the amines exerted their modulatory effect by one of the three mechanisms. First, they could compete for the ACh binding sites on the nAChR.[64] Second they could attach to a unique site that allosterically altered the effectiveness with which agonists operate the ion channel.[65-69] Third the amines could act at a receptor that is completely separate from the nAChR. This was suggested by the finding that the effect of 5-HT on ACh responses was greatly reduced following intracellular application of GDP-β-S indicating a possible indirect modulation of ACh responses.[57]

Conclusion

The functional data obtained from electrophysiological studies show that a simple description of nAChR subtypes from α-Bgt-sensitive and –insensitive are not sufficient to describe all the nAChR subtypes present in the insect CNS. In fact, in vertebrates, α-Bgt-sensitive receptors are shown to be highly permeable to Ca^{2+} while the second class groups heteromeric, α-Bgt-insensitive showed lower Ca^{2+} permeability. An extensive analysis of the Ca^{2+} permeability of insect native nAChRs is still lacking, mainly because neurons can express multiple nAChR subunits, yielding receptors of unknown composition. In addition, considering that distinct nAChR subtypes are present in the insect, recent findings suggested that α-Bgt-sensitive nAChRs could be also heteromeric receptors which highlight the complex subunit composition of native insect nAChRs.

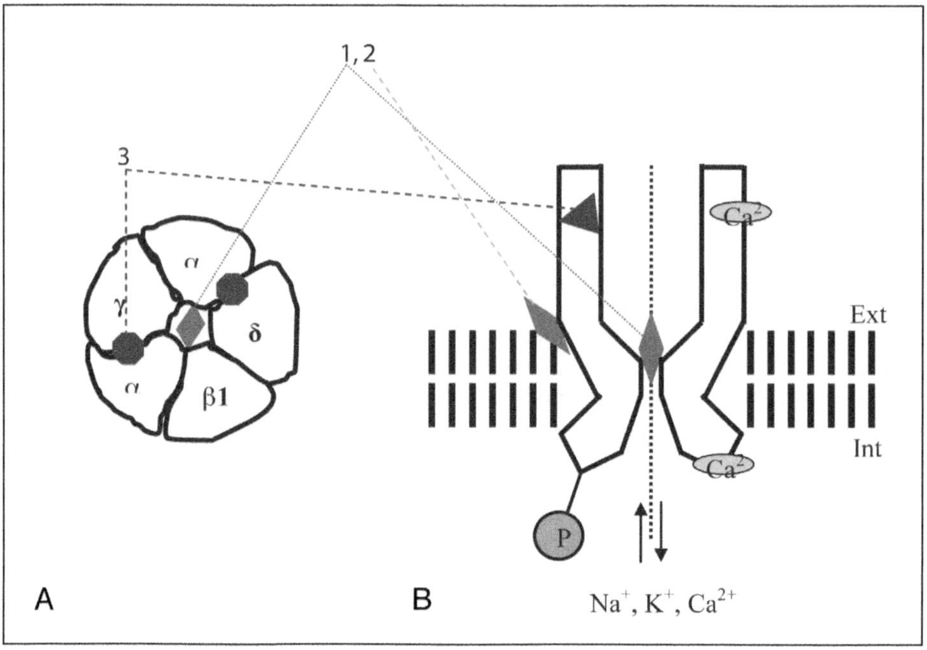

Figure 2. Schematic section of nicotinic receptors from vertebrate studies showing possible binding sites. (1) luminal non competitive inhibitors (i.g. chlorpromazine). (2) Non-luminal noncompetitive inhibitor binding sites. (3) ACh binding site and multiple allosteric sites including the noncompetitive allosteric activator site throughout the extracellular domain. Putative binding sites for Ca^{2+} at extracellular and intracellular domains were represented. Ext: extracellular domain; Int: intracellular domain.

Up to now, only sequences of cDNAs encoding calmodulin and CaMKII have been characterized on cockroach *Blatella germanica*[70] and *Periplaneta americana* (Unpublished data). Partial cDNA of Apis CaMKII was cloned and its expression pattern in the brain was demonstrated by *in situ* hybridization.[71] Nevertheless, there was not direct correlation between its expression and nicotinic receptors function.

Results from cockroach DUM neurons suggested that as vertebrate neuronal nAChRs,[72,73] phosphorylation plays a key role in the regulation of insect nAChRs.[1] This function may occur through consensus sequence (Fig. 2), in particular between the transmembrane domain TM3-TM4. In fact, most of the intracellular receptor portion is formed by this long cytoplasmic loop.[74] In the human α7 neuronal nAChR, mutation of conserved tyrosine 386 and 442 by alanine account for receptor insensitivity to kinase or phosphatase inhibition.[72] These tyrosines conserved in vertebrate nAChR subunits are not conserved in all insect nAChR subunits which may explain in part the different degree of phosphorylation between vertebrate and insect nAChRs. But, some caution are needed because it was known that mutations in the 'intermediate ring' of negatively charged residues, located at the cytoplasmic end of M2, reduced Ca^{2+} permeability without significantly modifying other functional properties such as activation and desensitization of the receptor.[75]

References

1. Thany SH, Lenaers G, Raymond-Delpech V et al. Exploring the pharmacological properties of insect nicotinic acetylcholine receptors. Trends Pharmacol Sci 2007; 28(1):14-22.
2. Goldberg F, Grunewald B, Rosenboom H et al. Nicotinic acetylcholine currents of cultured Kenyon cells from the mushroom bodies of the honey bee Apis mellifera. J Physiol 1999; 514(Pt 3):759-68.
3. Deglise P, Grunewald B, Gauthier M. The insecticide imidacloprid is a partial agonist of the nicotinic receptor of honeybee Kenyon cells. Neurosci Lett 2002; 321(1-2):13-6.
4. Wustenberg DG, Grunewald B. Pharmacology of the neuronal nicotinic acetylcholine receptor of cultured kenyon cells of the honeybee, Apis mellifera. J Comp Physiol A Neuroethol Sens Neural Behav Physiol 2004; 190(10):807-21.
5. Barbara GS, Grunewald B, Paute S et al. Study of nicotinic acetylcholine receptors on cultured antennal lobe neurones from adult honeybee brains. Invert Neurosci 2008; 8(1):19-29.
6. Lapied B, Le Corronc H, Hue B. Sensitive nicotinic and mixed nicotinic-muscarinic receptors in insect neurosecretory cells. Brain Res 1990; 533(1):132-6.
7. Courjaret R, Lapied B. Complex intracellular messenger pathways regulate one type of neuronal alpha-bungarotoxin-resistant nicotinic acetylcholine receptors expressed in insect neurosecretory cells (dorsal unpaired median neurons). Mol Pharmacol 2001; 60(1):80-91.
8. Courjaret R, Grolleau F, Lapied B. Two distinct calcium-sensitive and -insensitive pkc up- and down-regulate an alpha-bungarotoxin-resistant nAChR1 in insect neurosecretory cells (DUM neurons). Eur J Neurosci 2003; 17(10):2023-34.
9. Thany SH, Courjaret R, Lapied B. Effect of calcium on nicotine-induced current expressed by an atypical alpha-bungarotoxin-insensitive nAChR2. Neurosci Lett 2008; 438(3):317-21.
10. Thany SH. Agonist actions of clothianidin on synaptic and extrasynaptic nicotinic acetylcholine receptors expressed on cockroach sixth abdominal ganglion. Neurotoxicology 2009 (In press).
11. Cayre M, Buckingham SD, Yagodin S et al. Cultured insect mushroom body neurons express functional receptors for acetylcholine, gaba, glutamate, octopamine and dopamine. J Neurophysiol 1999; 81(1):1-14.
12. Su H, O'Dowd DK. Fast synaptic currents in drosophila mushroom body Kenyon cells are mediated by alpha-bungarotoxin-sensitive nicotinic acetylcholine receptors and picrotoxin-sensitive gaba receptors. J Neurosci 2003; 23(27):9246-53.
13. Rohrbough J, Broadie K. Electrophysiological analysis of synaptic transmission in central neurons of drosophila larvae. J Neurophysiol 2002; 88(2):847-60.
14. Wegener C, Hamasaka Y, Nassel DR. Acetylcholine increases intracellular ca^{2+} via nicotinic receptors in cultured pdf-containing clock neurons of drosophila. J Neurophysiol 2004; 91(2):912-23.
15. Hermsen B, Stetzer E, Thees R et al. Neuronal nicotinic receptors in the locust Locusta migratoria. Cloning and expression. J Biol Chem 1998; 273(29):18394-404.
16. Guez D, Belzunces LP, Maleszka R. Effects of imidacloprid metabolites on habituation in honeybees suggest the existence of two subtypes of nicotinic receptors differentially expressed during adult development. Pharmacol Biochem Behav 2003; 75(1):217-22.
17. Dacher M, Lagarrigue A, Gauthier M. Antennal tactile learning in the honeybee: Effect of nicotinic antagonists on memory dynamics. Neuroscience 2005; 130(1):37-50.
18. Fayyazuddin A, Zaheer MA, Hiesinger PR et al. The nicotinic acetylcholine receptor dalpha7 is required for an escape behavior in drosophila. PLoS Biol 2006; 4(3):e63.
19. Gauthier M, Dacher M, Thany SH et al. Involvement of alpha-bungarotoxin-sensitive nicotinic receptors in long-term memory formation in the honeybee (Apis mellifera). Neurobiol Learn Mem 2006; 86(2):164-74.
20. Thany SH, Crozatier M, Raymond-Delpech V et al. Apisalpha2, apisalpha7-1 and apisalpha7-2: Three new neuronal nicotinic acetylcholine receptor alpha-subunits in the honeybee brain. Gene 2005; 344:125-32.
21. Jones AK, Sattelle DB. The cys-loop ligand-gated ion channel gene superfamily of the red flour beetle, Tribolium castaneum. BMC Genomics 2007; 8:327.
22. Chamaon K, Schulz R, Smalla KH et al. Neuronal nicotinic acetylcholine receptors of Drosophila melanogaster: The alpha-subunit dalpha3 and the beta-type subunit ard co-assemble within the same receptor complex. FEBS Lett 2000; 482(3):189-92.
23. Chamaon K, Smalla KH, Thomas U et al. Nicotinic acetylcholine receptors of drosophila: Three subunits encoded by genomically linked genes can co-assemble into the same receptor complex. J Neurochem 2002; 80(1):149-57.

24. Vermehren A, Qazi S, Trimmer BA. The nicotinic alpha subunit mara1 is necessary for cholinergic evoked calcium transients in manduca neurons. Neurosci Lett 2001; 313(3):113-6.
25. Salgado VL, Saar R. Desensitizing and nondesensitizing subtypes of alpha-bungarotoxin-sensitive nicotinic acetylcholine receptors in cockroach neurons. J Insect Physiol 2004; 50(10):867-79.
26. Nauen R, Ebbinghaus-Kintscher U, Schmuck R. Toxicity and nicotinic acetylcholine receptor interaction of imidacloprid and its metabolites in Apis mellifera (hymenoptera: Apidae). Pest Manag Sci 2001; 57(7):577-86.
27. Barbara GS, Zube C, Rybak J et al. Acetylcholine, gaba and glutamate induce ionic currents in cultured antennal lobe neurons of the honeybee, Apis mellifera. J Comp Physiol A Neuroethol Sens Neural Behav Physiol 2005; 191(9):823-36.
28. David JA, Pitman RM. The pharmacology of alpha-bungarotoxin-resistant acetylcholine rececptors on an identified cockroach motoneurone. J Comp Physiol [A] 1993; 172:359-368.
29. Hancox JC, Pitman RM. Plateau potentials drive axonal impulse burst in insect motoneurons. Proc R Soc Lond B 1991; 244:33-38.
30. Hancox JC, Pitman RM. A time-dependent excitability change in the soma of an identified insect motoneuron. J Exp Biol 1992; 162:251-263.
31. Hancox JC, Pitman RM. Plateau potentials in an insect motoneuron can be driven by synaptic stimulation. J Exp Biol 1993; 176:307-310.
32. Mills JD, Pitman RM. Electrical properties of a cockroach motor neuron soma depend on different characteristics of individual ca components. J Neurophysiol 1997; 78(5):2455-66.
33. Mills JD, Pitman RM. Contribution of potassium conductances to a time-dependent transition in electrical properties of a cockroach motoneuron soma. J Neurophysiol 1999; 81(5):2253-66.
34. Benson JA. Electrophysiological pharmacology of the nicotinic and muscarinic cholinergic responses of isolated neuronal somata from locust thoracic ganglia. J Exp Biol 1992; 170:203-233.
35. Elgoyhen AB, Johnson DS, Boulter J et al. Alpha 9: An acetylcholine receptor with novel pharmacological properties expressed in rat cochlear hair cells. Cell 1994; 79(4):705-15.
36. Verbitsky M, Rothlin CV, Katz E et al. Mixed nicotinic-muscarinic properties of the alpha9 nicotinic cholinergic receptor. Neuropharmacology 2000; 39(13):2515-24.
37. Elgoyhen AB, Vetter DE, Katz E et al. Alpha10: A determinant of nicotinic cholinergic receptor function in mammalian vestibular and cochlear mechanosensory hair cells. Proc Natl Acad Sci USA 2001; 98(6):3501-6.
38. Hiel H, Elgoyhen AB, Drescher DG et al. Expression of nicotinic acetylcholine receptor mrna in the adult rat peripheral vestibular system. Brain Res 1996; 738(2):347-52.
39. Morley BJ, Li HS, Hiel H et al. Identification of the subunits of the nicotinic cholinergic receptors in the rat cochlea using RT-PCR and in situ hybridization. Brain Res Mol Brain Res 1998; 53(1-2):78-87.
40. Changeux JP, Bertrand D, Corringer PJ et al. Brain nicotinic receptors: Structure and regulation, role in learning and reinforcement. Brain Res Brain Res Rev 1998; 26(2-3):198-216.
41. Paterson D, Nordberg A. Neuronal nicotinic receptors in the human brain. Prog Neurobiol 2000; 61(1):75-111.
42. Oertner TG, Single S, Borst A. Separation of voltage- and ligand-gated calcium influx in locust neurons by optical imaging. Neurosci Lett 1999; 274(2):95-8.
43. Fucile S, Sucapane A, Eusebi F. Ca^{2+} permeability through rat cloned alpha9-containing nicotinic acetylcholine receptors. Cell Calcium 2006; 39(4):349-55.
44. Dajas-Bailador F, Wonnacott S. Nicotinic acetylcholine receptors and the regulation of neuronal signalling. Trends Pharmacol Sci 2004; 25(6):317-24.
45. Vermehren A, Trimmer BA. Expression and function of two nicotinic subunits in insect neurons. J Neurobiol 2005; 62(3):289-98.
46. Grolleau F, Lapied B, Buckingham SD et al. Nicotine increases $[ca^{2+}]_i$ and regulates electrical activity in insect neurosecretory cells (DUM neurons) via an acetylcholine receptor with 'mixed' nicotinic-muscarinic pharmacology. Neurosci Lett 1996; 220(2):142-6.
47. Heine M, Wicher D. Ca^{2+} resting current and Ca^{2+}-induced Ca^{2+} release in insect neurosecretory neurons. Neuroreport 1998; 9(14):3309-14.
48. Wicher D, Messutat S, Lavialle C et al. A new regulation of noncapacitative calcium entry in insect pacemaker neurosecretory neurons. Involvement of arachidonic acid, no-guanylyl cyclase/cGMP and cAMP. J Biol Chem 2004; 279(48):50410-9.
49. Berridge MJ, Lipp P, Bootman MD. The versatility and universality of calcium signalling. Nat Rev Mol Cell Biol 2000; 1(1):11-21.

50. Newton AC. Protein kinase C. Seeing two domains. Curr Biol 1995; 5(9):973-6.
51. Newton AC. Protein kinase C: Structure, function and regulation. J Biol Chem 1995; 270(48):28495-8.
52. Newton AC. Protein kinase C: Structural and spatial regulation by phosphorylation, cofactors and macromolecular interactions. Chem Rev 2001; 101(8):2353-64.
53. Hou J, Kuromi H, Fukasawa Y et al. Repetitive exposures to nicotine induce a hyper-responsiveness via the cAMP/PKA/CREB signal pathway in drosophila. J Neurobiol 2004; 60(2):249-61.
54. Dymond GR, Evans PD. Biogenic amines in the nervous system of the cockroach, Periplaneta americana: Association of octopamine with mushroom bodies and dorsal unpaired median (DUM) neurones. Insect Biochem 1979; 9:535-545.
55. Sloley BD, Owen MD. The effects of reserpine on amine concentrations in the nervous system of the cockroach (Periplaneta americana). Insect Biochem 1982; 12:469-476.
56. Shafi N, Midgley JM, Matson DG et al. Analysis of biogenic-amines in the brain of the american cockroach (Periplaneta americana) by gas-chromatography negative-ion chemical ionization mass-spectrometry. J Chromatog Biomed Applications 1989; 490:9-19.
57. Butt SJ, Pitman RM. Modulation by 5-hydroxytryptamine of nicotinic acetylcholine responses recorded from an identified cockroach (Periplaneta americana) motoneuron. Eur J Neurosci 2002; 15(3):429-38.
58. Butt SJ, Pitman RM. Indirect phosphorylation-dependent modulation of postsynaptic nicotinic acetylcholine responses by 5-hydroxytryptamine. Eur J Neurosci 2005; 21(5):1181-8.
59. Butt SJB, Pitman RM. Modulation by monoamines of ACh responses recorded from an identified cockroach (Periplaneta americana) motoneurone. J Physiol (London) 1998; 513:102P.
60. Jackson C, Bermudez I, Beadle DJ. Pharmacological properties of nicotinic acetylcholine receptors in isolated locusta migratoria neurones. Microsc Res Tech 2002; 56(4):249-55.
61. Benson JA. Bicuculline blocks the response to acetylcholine and nicotine but not to muscarinie or gaba in isolated insect neuronal somata. Brain Res 1988; 458:45-71.
62. Marshall J, Buckingham SD, Shingai R et al. Sequence and functional expression of a single alpha subunit of an insect nicotinic acetylcholine receptor. EMBO J 1990; 9(13):4391-8.
63. Buckingham SD, Hue B, Sattelle DB. Actions of bicuculline on cell body and neuropilar membranes of identified insect neurones. J Exp Biol 1994; 186:235-44.
64. Akasu T, Koketsu K. 5-hydroxytryptamine decrease the sensitivity of nicotinic acetylcholine receptors in bull-frog sympathetic ganglion. J Physiol 1986; 380:93-109.
65. Grassi F, Polenzani L, Mileo AM et al. Blockage of nicotinic acetylcholine receptors by 5-hydroxytryptamine. J Neurosci Res 1993; 34(5):562-70.
66. Schrattenholz A, Pereira EF, Roth U et al. Agonist responses of neuronal nicotinic acetylcholine receptors are potentiated by a novel class of allosterically acting ligands. Mol Pharmacol 1996; 49(1):1-6.
67. Garcia-Colunga J, Miledi R. Blockage of mouse muscle nicotinic receptors by serotonergic compounds. Exp Physiol 1999; 84(5):847-64.
68. Nakazawa K, Ohno Y. Block by 5-hydroxytryptamine and apomorphine of recombinant human neuronal nicotinic receptors. Eur J Pharmacol 1999; 374(2):293-9.
69. Blanton MP, McCardy EA, Fryer JD et al. 5-hydroxytryptamine interaction with the nicotinic acetylcholine receptor. Eur J Pharmacol 2000; 389(2-3):155-63.
70. Inagaki S, Kaku K, Dunlap DY et al. Sequences of cDNAs encoding calmodulin and partial structures of calmodulin kinase and a calcium channel of kdr-resistant and -susceptible german cockroaches, Blattella germanica. Comp Biochem Physiol C Pharmacol Toxicol Endocrinol 1998; 120(2):225-33.
71. Kamikouchi A, Takeuchi H, Sawata M et al. Concentrated expression of Ca^{2+}/calmodulin-dependent protein kinase II and protein kinase C in the mushroom bodies of the brain of the honeybee Apis mellifera. J Comp Neurol 2000; 417(4):501-10.
72. Charpantier E, Wiesner A, Huh KH et al. Alpha7 neuronal nicotinic acetylcholine receptors are negatively regulated by tyrosine phosphorylation and SRC-family kinases. J Neurosci 2005; 25(43):9836-49.
73. Marszalec W, Yeh JZ, Narahashi T. Desensitization of nicotine acetylcholine receptors: Modulation by kinase activation and phosphatase inhibition. Eur J Pharmacol 2005; 514(2-3):83-90.
74. Kuo YP, Xu L, Eaton JB et al. Roles for nicotinic acetylcholine receptor subunit large cytoplasmic loop sequences in receptor expression and function. J Pharmacol Exp Ther 2005; 314(1):455-66.
75. Bertrand D, Galzi JL, Devillers-Thiéry A et al. Mutations at two distinct sites within the channel domain M2 alter calcium permeability of neuronal alpha7 nicotinic receptor. Proc Natl Acad Sci USA 1993; 90(1):6971-6975.

CHAPTER 6

Characterisation of Insect Nicotinic Acetylcholine Receptors by Heterologous Expression

Neil S. Millar* and Stuart J. Lansdell

Abstract

As with other neurotransmitter receptors and ion channels, characterisation of nicotinic acetylcholine receptors (nAChRs) has relied heavily on studies conducted with cloned receptors expressed in artificial expression systems. Although much has been achieved in recent years by such studies, considerable problems have been encountered in the heterologous expression of several nAChR subtypes; problems that have been particularly pronounced for insect nAChRs. Here we will review studies that have been conducted with nAChRs cloned from insects, with emphasis on experimental strategies that have been employed in an attempt to circumvent the problems associated with inefficient heterologous expression of insect nAChRs. These approaches include the expression of hybrid nAChRs (containing insect nAChR subunit co-expressed with vertebrate subunits), artificial subunit chimeras and the co-expression of molecular chaperones such as RIC-3.

Introduction

Nicotinic acetylcholine receptors (nAChRs) are members of the super-family of ligand-gated ion channels. In insects, as in vertebrate species, nAChRs are a major subtype of excitatory neurotransmitter receptor. Nicotinic receptors are the most abundant excitatory postsynaptic receptors in insects,[1] where they play an important role in synaptic transmission and are also a major target site for commercially important insecticides.[2,3]

Nicotinic receptors are complex transmembrane proteins in which five subunits co-assemble to form an oligomeric complex with a central cation-selective pore.[4] In vertebrate species, where nAChRs are expressed both within the nervous system and at the neuromuscular junction, seventeen distinct subunits have been identified (α1-α10, β1-β4, γ, δ and ε).[5,6] In most insect species that have been examined by genome sequencing about ten nAChR subunits have been identified.[6-11] As has been discussed in detail elsewhere,[3,6] the nomenclature of insect nAChR subunits is somewhat inconsistent amongst different species. In the fruit fly *Drosophila melanogaster*, the insect for which nAChRs have been studied in greatest detail, the ten nAChR subunits are commonly referred to as Dα1-Dα7 and Dβ1-Dβ3.[3,6] However, an alternative subunit nomenclature has also been used extensively in the literature for *Drosophila* nAChR subunits, for example ALS, SAD, ARD and SBD (for Dα1, Dα2, Dβ1 and Dβ2, respectively). As with vertebrate nAChRs, the convention[12]

*Corresponding Author: Neil S. Millar—Department of Neuroscience, Physiology and Pharmacology, University College London, London, WC1E 6BT, UK. Email: n.millar@ucl.ac.uk

Insect Nicotinic Acetylcholine Receptors, edited by Steeve Hervé Thany.

is for nAChR subunits to be referred to as α-type subunits if they contain two adjacent Cys amino acids at positions equivalent to Cys192 and 193 in the nAChR α subunit from the electric organ of the marine ray *Torpedo*.[13] The original assumption was that α subunits were 'agonist binding' subunits, whereas non-α subunits were structural subunits. However, it is now widely accepted that agonists bind at subunit interfaces and, typically, at interfaces between an α and a non-α subunit. Some nAChRs subunits (always α subunits) are capable of forming functional homomeric receptors; the most extensively studied example being the vertebrate α7 subunit.[14] However, more commonly, nAChRs are heteromeric complexes containing both α and non-α subunits.

Characterization of Insect nAChRs by Heterologous Expression

There have been very few reports of the functional expression of insect nAChRs other than studies of 'hybrid' receptors in which insect nAChR subunits have been co-expressed with vertebrate nAChR subunits (Table 1; see below for details of studies with hybrid receptors). One

Table 1. Heterologous expression of insect nAChRs

Species	Common Name	Subunit (Alternative Nomenclature)	References*
Ctenocephalides Felis	Cat flea	Cfα1	29
		Cfα3	29
Drosophila Melanogaster	Fruit fly	Dα1 (ALS)	19, 31, 33, 61, 65, 67
		Dα2 (SAD)	18, 19, 31, 33, 59, 61, 64-70, 73, 74
		Dα3	33, 40, 73
		Dα4	34, 73
		Dα6	53
		Dα7	53
		Dβ1 (ARD)	73
		Dβ2 (SBD)	31, 73
		Dβ3	73
Myzus persicae	Peach potato aphid	Mpα1	20, 32, 41
		Mpα2	20, 32, 41, 74, 75
		Mpα3	32
		Mpβ1	41
Nilaparvata lugens	Brown planthopper	Nlα1	28, 35, 71, 72
		Nlα2	63
		Nlα3	42
Schistocerca gregaria	Desert locust	Sgα1 (αL1)	15-17

*References cited are those that report studies in which insect nAChR subunits have been examined by heterologous expression (specifically, expression of cloned subunits in either *Xenopus* oocytes or cultured cell lines). As is discussed in the text, in almost all cases, such studies have involved co-expression of non-insect nAChR subunits (hybrid receptors) or the construction of subunit chimeras. Cited references are restricted to those in which evidence of successful heterologous expression has been presented (for example, by electrophysiological, radioligand binding or co-immunoprecipitation data). As is discussed in the text, nAChR subunits have been cloned from several other insect species for which no data from heterologous expression studies has been described. These include nAChR subunits cloned from the honeybee *Apis mellifera*,[21,22] the house fly *Musca domestica*,[23-25] the locust *Locusta migratoria*,[30] the red flour beetle *Tribolium castaneum*,[26] the silkworm *Bombyx mori*[11] and the tobacco hornworm *Manduca sexta*.[27]

of the few examples of successful functional expression of insect recombinant nAChRs is the expression in *Xenopus* oocytes of a nAChR α subunit (Sgα1, also referred to as αL1) cloned from the locust *Schistocerca gregaria*.[15-17] When this nAChR subunit is expressed alone in *Xenopus* oocytes, dose-dependent whole-cell responses to agonists can be detected which are blocked by nicotinic antagonists such as α-bungarotoxin, d-tubocurarine and methyllycaconitine.[15-17]

There have been reports of the functional expression in *Xenopus* oocytes of homomeric nAChRs containing the Dα2 (SAD) subunit from the fruit fly *Drosophila melanogaster*, although agonist-evoked responses were detected only with very high agonist concentrations.[18] In contrast, other studies have failed to detect functional nAChRs in *Xenopus* oocytes when Dα2 is expressed alone.[19] The Mpα1 and Mpα2 subunits, cloned from the aphid *Myzus persicae*, have also been reported to generate functional homomeric nAChRs in *Xenopus* oocytes,[20] although agonist-evoked responses were small and were expressed inefficiently.[20]

The cloning of nAChR subunits has been reported from several insect species. However, in most cases successful functional expression has either not been reported (for example with nAChRs cloned from the honeybee *Apis mellifera*,[21,22] the house fly *Musca domestica*,[23-25] the red flour beetle *Tribolium castaneum*,[26] the silkworm Bombyx mori[11] and the tobacco hornworm *Manduca sexta*[27]), or attempts to generate functional recombinant nAChRs have been reported as being unsuccessful (for example, with nAChR subunits cloned from the brown planthopper *Nilaparvata lugens*,[28] the cat flea *Ctenocephalides felis*[29] and the locust *Locusta migratoria*,[30]). Thus, there have been only very limited examples of the successful heterologous expression of functional insect nAChRs and, in all cases where functional expression has been reported, this has been achieved with the *Xenopus* oocyte expression system. Attempts have been made to express insect nAChRs cloned from several insect species in a variety of cultured cell lines, including both mammalian and insect cell lines, but so far these approaches have proved to be unsuccessful.[29,31-35] This is both puzzling and frustrating, particularly since other insect ligand-gated ion channels (such as the GABA-gated RDL receptor from *Drosophila*) generate functional recombinant receptors readily in a variety of cultured cell lines.[36-38]

As is discussed below, despite the problems that have been encountered in the heterologous expression of insect nAChRs, several insect α subunits have been successfully co-expressed in artificial expression systems as hybrid recombinant nAChRs, in which insect α subunits are co-assembled with vertebrate non-α subunits.[19,28,31,32,35] It was widely assumed that these difficulties might be a consequence of a requirement for an as-yet unidentified insect non-α subunit.[39] However, with the completion of several insect genome projects, this possibility seems less likely. Another possibility is that insect nAChRs have a requirement for specific accessory proteins or molecular chaperones. Indeed, as will be discussed later, several recent lines of evidence have provided support for this latter conclusion.

Characterization of Hybrid nAChRs

Problems associated with the heterologous expression of insect nAChRs (as described above) have prompted the use of several experimental strategies aimed at overcoming these problems. One of the most successful approaches has been to generate hybrid recombinant nAChRs by the co-expression of insect α subunits with vertebrate non-α subunits. This approach has permitted the functional expression in *Xenopus* oocytes of hybrid nAChRs containing a *Drosophila* nAChR α subunit (Dα1, Dα2 or Dα3) co-assembled with the chicken nAChR β2 subunit.[19,40] A similar approach, in which the *Drosophila* Dα1, Dα2, Dα3 or Dα4 subunits were co-expressed with the rat β2 subunit, has enabled the expression of hybrid nAChRs in cultured cell lines.[31,33,34] Although functional expression of these hybrid receptors (by electrophysiological techniques) has not been reported in cultured cell lines, high affinity binding of nicotinic radioligands has been detected and has permitted the pharmacological properties of these hybrid nAChRs to be examined.[31,33,34] In addition, studies with *Drosophila* nAChR α subunits have demonstrated that hybrid nAChRs can be formed by co-expression with a variety of vertebrate non-α subunits (β2, β4, γ and δ) and has helped to demonstrate the influence of subunit composition upon pharmacological properties of nAChRs.[31,33] A similar approach (in which insect α subunits are co-expressed with the vertebrate β2 subunit) has been used successfully for the characterization of hybrid nAChRs containing α subunits cloned from a variety of insect species. This has included nAChR α subunits cloned from the aphid *Myzus persicae*,[32,41] the brown plant hopper *Nilaparvata lugens*[28,35,42] and the cat flea *Ctenocephalides felis*.[29] Where such studies have been performed in *Xenopus* oocytes (see, for example, refs. 19, 40), functional expression of hybrid nAChRs has been observed. In contrast, where studies have been performed in cultured cell lines (see, for example, refs. 31, 33), hybrid nAChRs have been characterised by radioligand binding.

Interestingly, although high affinity binding of nicotinic radioligands can be detected with hybrid (*Drosophila*/rat) nAChRs in a *Drosophila* cell line maintained at 25°C, specific binding in mammalian cells (which are normally maintained at 37°C) has been detected only after mammalian cells are cultured at a lower temperature.[31] A plausible explanation for this observation appears to be that insect nAChR subunits are unable to fold efficiently at 37°C,[31] an observation which has been made previously for nAChRs cloned from the marine ray *Torpedo*.[43]

Characterization of Chimeric nAChR Subunits

An alternative strategy, aimed at circumventing problems associated with heterologous expression of insect nAChRs, is the construction and expression of nAChR subunit chimeras. This is an approach which was originally pioneered with the vertebrate nAChR α7 subunit.[44] The nAChR α7 subunit efficiently forms functional homomeric nAChRs when expressed in *Xenopus* oocytes,[14] but fails to do so in many cultured mammalian cell lines, where neither functional expression nor specific radioligand binding can be detected.[45-47] Despite these difficulties in the expression of α7 nAChRs in mammalian cells, a subunit chimera containing the N-terminal domain of the α7 subunit fused to the C-terminal domain of the mouse 5-HT_3 receptor 3A subunit has been shown to generate functional receptors very efficiently in a wide range of mammalian cell types,[44,45] a finding which has been extended by the construction of similar subunit chimeras containing the N-terminal domain of other vertebrate nAChR subunits.[48-52]

Expression studies with the *Drosophila* Dα6 and Dα7 subunits in a cultured *Drosophila* cell line failed to generate nAChRs which could be detected by radioligand binding.[53] In addition, co-expression of Dα6 and Dα7 with vertebrate non-α subunits (an approach that had been successful with the Dα1-Dα4 subunits[31,33,34]) also failed to alleviate this problem. However, construction of subunit chimeras in which the N-terminal region of either Dα6 and Dα7 were fused to the C-terminal domain of the 5-HT_3 receptor 3A subunit, results in the formation of chimeric subunits that are able to assemble into complexes that can be characterized by radioligand binding techniques.[34]

In some cases, an even more convoluted approach has been adopted to characterize insect nAChR subunits, for example, combining both artificial subunit chimeras with the co-expression with non-insect nAChR subunits. In cases where only partial insect nAChR α subunit cDNA sequences were available, for example in the case of subunits cloned from the cat flea, subunit chimeras have been constructed in which the N-terminal ligand binding domain of cat flea nAChR subunits were fused to the C-terminal domain of a *Drosophila* α subunit.[29] Co-expression of these insect α subunit chimeras, together with the rat β2 subunit, has been employed to facilitate pharmacological characterization of these chimeric and hybrid nAChRs by radioligand binding.[29]

Co-Expression of Molecular Chaperones

There is increasing evidence that some of the problems associated with heterologous expression of insect (and also of vertebrate) nAChRs may be a consequence of a requirement for molecular chaperones such as the recently identified protein RIC-3.[54] RIC-3 is a transmembrane protein that was originally identified in the nematode *Caenorhabditis elegans.*[55] It has been shown to enhance levels of functional expression of vertebrate nAChR, such as homomeric α7 receptors in *Xenopus* oocytes[55] but, perhaps of even greater significance, has been the finding that RIC-3 is able to facilitate functional expression of α7 nAChRs in mammalian cell lines which are otherwise unable to support functional expression of α7 nAChRs.[56-58] Such findings have prompted studies to examine whether RIC-3 might be required for the successful functional expression of insect nAChRs. In a recent study, several alternatively spliced RIC-3 homologues were cloned from *Drosophila* and their nAChR chaperone activity compared with that of human RIC-3.[59] It has been shown that human RIC-3 is a more efficient nAChR chaperone in human cell lines, whereas *Drosophila* RIC-3 is a more efficient chaperone in a *Drosophila* cell line.[59] However, although co-expression of RIC-3 has been found to enhance levels of hybrid (insect/mammalian) nAChRs, as assayed by radioligand binding, as yet, RIC-3 has not been shown to facilitate heterologous expression of insect-only recombinant nAChRs.

Pharmacological Properties of Recombinant nAChRs

The problems described above (concerning heterologous expression of insect nAChRs) have limited the extent to which the properties of insect nAChRs can be examined by heterologous expression. In particular, the subunit composition of native insect nAChRs remains unclear. However, co-immunoprecipitation studies with native nAChRs expressed in *Drosophila* have provided some insights into this question.[60-62] Despite the difficulties associated with heterologous expression, studies conducted with recombinant nAChRs have also provided some information about co-assembly of insect nAChR subunits. For example, electrophysiological studies in *Xenopus* oocytes from have provided evidence that two different insect nAChR α subunits can co-assemble within the same receptor complex.[61,63]

A particular focus of studies with insect recombinant nAChRs has been to examine the influence of subunit composition and of subunit domains upon neonicotinoid insecticides such as imidacloprid. Electrophysiological and radioligand binding studies conducted with hybrid nAChRs have identified subunits that contribute to neonicotinoid-sensitive receptors. Subunits identified by these approaches include Dα1-Dα3 from *Drosophila*,[33,64-66] Mpα2 and Mpα3 from *M. persicae*[32] and also Nlα1 and Nlα2 from *N. lugens*.[28,35,63] In addition, such studies have demonstrated the contribution of specific insect nAChRs subunit in generating receptors sensitive to naturally occurring toxins such as α-bungarotoxin[33,53] and nereistoxin.[67] Techniques such as site-directed mutagenesis and construction of subunit chimeras have also been exploited in identifying subunit domains and individual amino acids which are important in the binding of neonicotinoid insecticides to insect nAChRs.[42,68-72]

Conclusion

As has been discussed above, many of the difficulties that have been encountered in the heterologous expression of insect nAChRs are still unresolved. To an extent, these problems can be circumvented by expression of either hybrid nAChRs or subunit chimeras. Indeed, such approaches have provided an important insight into the pharmacological properties of insect nAChRs and into the relationship between the structural and functional properties of these receptors. However, extrapolation from recent advances with vertebrate nAChRs suggests that efficient functional expression of insect recombinant nAChRs may require a better understanding of the role played by the host cell and, in particular, of nAChR-associated chaperone proteins.

References

1. Sattelle DB. Acetylcholine receptors of insects. Adv Insect Physiol 1980; 15:215-315.
2. Matsuda K, Buckingham SD, Kleier D et al. Neonicotinoids: insecticides acting on insect nicotinic acetylcholine receptors. Trends Pharm Sci 2001; 22:573-580.
3. Millar NS, Denholm I. Nicotinic acetylcholine receptors: targets for commercially important insecticides. Invert Neurosci 2007; 7:53-66.
4. Green WN, Millar NS. Ion-channel assembly. Trends Neurosci 1995; 18:280-287.
5. Le Novère N, Corringer P-J, Changeux J-P. The diversity of subunit composition in nAChRs: evolutionary origins, physiologic and pharmacologic consequences. J Neurobiol 2002; 53:447-456.
6. Millar NS. Assembly and subunit diversity of nicotinic acetylcholine receptors. Biochem Soc Trans 2003; 31:869-874.
7. Jones AK, Brown AM, Sattelle DB. Insect nicotinic acetylcholine receptor gene families: from genetic model organisms to vector, pest and beneficial species. Invert Neurosci 2007; 7:67-73.
8. Jones AK, Grauso M, Sattelle DB. The nicotinic acetylcholine receptor gene family of the malaria mosquito, Anopheles gambiae. Genomics 2005; 85:176-187.
9. Jones AK, Raymond-Delpech V, Thany SH et al. The nicotinic acetylcholine receptor gene family of the honey bee, Apis mellifera. Genome Res 2006; 16:1422-1430.
10. Jones AK, Sattelle DB. The cys-loop ligand-gated ion channel gene superfamily of the red flour beetle, Tribolium castaneum. BMC Genomics 2007; 8:327 doi: 310.1186/1471-2164-1188-1327.
11. Shao Y-M, Dong K, Zhang C-X. The nicotinic acetylcholine receptor gene family of the silkworm, Bombyx mori. BMC Genomics 2007; 8:324 doi:310.1186/1471-2164-1188-1324.
12. Lukas RJ, Changeux J-P, Le Novère N et al. International union of pharmacology. XX. Current status of the nomenclature for nicotinic acetylcholine receptors and their subunits. Pharmacol Rev 1999; 51:397-401.
13. Noda M, Takahashi H, Tanabe T et al. Primary structure of α-subunit precursor of Torpedo californica acetylcholine receptor deduced from cDNA sequence. Nature 1982; 299:793-797.
14. Couturier S, Bertrand D, Matter JM et al. A neuronal nicotinic acetylcholine receptor subunit (α7) is developmentally regulated and forms a homo-oligomeric channel blocked by α-BTX. Neuron 1990; 5:847-856.
15. Marshall J, Buckingham SD, Shingai R et al. Sequence and functional expression of a single α subunit of an insect nicotinic acetylcholine receptor. EMBO J 1990; 9:4391-4398.
16. Amar M, Thomas P, Wonnacott S et al. A nicotinic acetylcholine receptor subunit from insect brain forms a nondesensitising homo-oligomeric nicotinic acetylcholine receptor when expressed in Xenopus oocytes. Neurosci Letts 1995; 199:107-110.
17. Jones AK, Marshall J, Blake AD et al. Sgβ1, a novel locust (Schistocerca gregaria) non-α nicotinic acetylcholine receptor-like subunit with homology to the Drosophila melanogaster Dβ1 subunit. Invert Neurosci 2005; 5:147-155.
18. Sawruk E, Schloss P, Betz H et al. Heterogeneity of Drosophila nicotinic acetylcholine receptors: SAD, a novel developmentally regulated α-subunit. EMBO J 1990; 9:2671-2677.
19. Bertrand D, Ballivet M, Gomez M et al. Physiological properties of neuronal nicotinic receptors reconstituted from the vertebrate β2 subunit and Drosophila α subunits. Eur J Neurosci 1994; 6:869-875.
20. Sgard F, Fraser SP, Katkowska MJ et al. Cloning and functional characterisation of two novel nicotinic acetylcholine receptor α subunits from the insect pest Myzus persicae. J Neurochem 1998; 71:903-912.
21. Thany SH, Lenaers G, Crozatier M et al. Identification and localization of the nicotinic acetylcholine receptor alpha3 mRNA in the brain of the honeybee, Apis mellifera. Insect Mol Biol 2003; 12:255-262.
22. Thany SH, Crozatier M, Raymond-Delpech V et al. Apisα2, Apisα7-1 and Apisα7-2: three new neuronal nicotinic acetylcholine receptor α-subunits in the honeybee brain. Gene 2005; 344:125-132.
23. Gao J-R, Deacutis JM, Scott JG. The nicotinic acetylcholine receptor subunit Mdα5 and Mdβ3 on autosome 1 of Musca domestica are not involved in spinosad resistance. Insect Mol Biol 2007; 16:691-701.
24. Gao J-R, Deacutis JM, Scott JG. Characterization of the nicotinic acetylcholine receptor gene Mdα2 from the house fly Musca domestica. Arch Insect Biochem Physiol 2007; 64:30-42.
25. Gao J-R, Deacutis JM, Scott JG. The nicotinic acetylcholine receptor subunit Mdα6 from Musca domestica is diversified via posttranscriptional modification. Insect Mol Biol 2007; 16:325-334.

26. Rinkevich FD, Scott JG. Transcriptional diversity and allelic variation in nicotinic acetylcholine receptor subunits of the red flour beetle, Tribolium castaneum. Insect Mol Biol 2009; 18:233-242.
27. Eastham HM, Lind RJ, Eastlake JL et al. Characterization of a nicotinic acetylcholine receptor from the insect Manduca sexta. Eur J Neurosci 1998; 10:879-889.
28. Liu Z, Williamson MS, Lansdell SJ et al. A nicotinic acetylcholine receptor mutation (Y151S) causes reduced agonist potency to a range of neonicotinoid insecticides. J Neurochem 2006; 99:1273-1281.
29. Bass C, Lansdell SJ, Millar NS et al. Molecular characterisation of nicotinic acetylcholine receptor subunits from the cat flea, Ctenocephalides felis (Siphonaptera: Pulicidae). Insect Biochem Mol Biol 2006; 36:86-96.
30. Hermsen B, Stetzer E, Thees R et al. Neuronal nicotinic receptors in the Locust Locusta migratoria: cloning and expression. J Biol Chem 1998; 273:18394-18404.
31. Lansdell SJ, Schmitt B, Betz H et al. Temperature-sensitive expression of Drosophila neuronal nicotinic acetylcholine receptors. J Neurochem 1997; 68:1812-1819.
32. Huang Y, Williamson MS, Devonshire AL et al. Molecular characterization and imidacloprid selectivity of nicotinic acetylcholine receptor subunits from the peach-potato aphid Myzus persicae. J Neurochem 1999; 73:380-389.
33. Lansdell SJ, Millar NS. The influence of nicotinic receptor subunit composition upon agonist, α-bungarotoxin and insecticide (imidacloprid) binding affinity. Neuropharmacol 2000; 39:671-679.
34. Lansdell SJ, Millar NS. Cloning and heterologous expression of Dα4, a Drosophila neuronal nicotinic acetylcholine receptor subunit: identification of an alternative exon influencing the efficiency of subunit assembly. Neuropharmacol 2000; 39:2604-2614.
35. Liu Z, Williamson MS, Lansdell SJ et al. A nicotinic acetylcholine receptor mutation conferring target-site resistance to imidacloprid in Nilaparvata lugens (brown planthopper). Proc Natl Acad Sci USA 2005; 102:8420-8425.
36. Lee H-J, Rocheleau T, Zhang H-G et al. Expression of a Drosophila GABA receptor in a baculovirus insect cell system. FEBS Lett 1993; 335:315-318.
37. Millar NS, Buckingham SD, Sattelle DB. Stable expression of a functional homo-oligomeric Drosophila GABA receptor in a Drosophila cell line. Proc R Soc Lond B 1994; 258:307-314.
38. Shotkoski F, Zhang HG, Jackson MB et al. Stable expression of insect GABA receptors in insect cell lines. Promoters for efficient expression of Drosophila and mosquito Rdl GABA receptors in stably transformed mosquito cell lines. FEBS Lett 1996; 380:257-262.
39. Millar NS. Heterologous expression of mammalian and insect neuronal nicotinic acetylcholine receptors in cultured cell lines. Biochem Soc Trans 1999; 27:944-950.
40. Schulz R, Sawruk E, Mülhardt C et al. Dα3, a new functional a subunit of nicotinic acetylcholine receptors from Drosophila. J Neurochem 1998; 71:853-862.
41. Huang Y, Williamson MS, Devonshire AL et al. Cloning, heterologous expression and co-assembly of Mpβ1, a nicotinic acetylcholine receptor subunit from the aphid Myzus persicae. Neurosci Lett 2000; 284:116-120.
42. Liu Z, Han Z, Liu S et al. Amino acids outside of the loops that define the agonist binding site are important for ligand binding to insect nicotinic acetylcholine receptors. J Neurochem 2008; 106:224-230.
43. Claudio T, Green WN, Hartman DS et al. Genetic reconstitution of functional acetylcholine receptor channels in mouse fibroblasts. Science 1987; 238:1688-1694.
44. Eiselé J-L, Bertrand S, Galzi J-L et al. Chimaeric nicotinic-serotonergic receptor combines distinct ligand binding and channel specificities. Nature 1993; 366:479-483.
45. Cooper ST, Millar NS. Host cell-specific folding and assembly of the neuronal nicotinic acetylcholine receptor α7 subunit. J Neurochem 1997; 68:2140-2151.
46. Kassner PD, Berg DK. Differences in the fate of neuronal acetylcholine receptor protein expressed in neurons and stably transfected cells. J Neurobiol 1997; 33:968-982.
47. Rangwala F, Drisdel RC, Rakhilin S et al. Neuronal α-bungarotoxin receptors differ structurally from other nicotinic acetylcholine receptors. J Neurosci 1997; 17:8201-8212.
48. Cooper ST, Millar NS. Host cell-specific folding of the neuronal nicotinic receptor α8 subunit. J Neurochem 1998; 70:2585-2593.
49. Quiram PA, Sine SM. Identification of residues in the neuronal α7 acetylcholine receptor that confer selectivity for conotoxin ImI. J Biol Chem 1998; 273:11001-11006.

50. Cooper ST, Harkness PC, Baker ER et al. Upregulation of cell-surface α4β2 neuronal nicotinic receptors by lower temperature and expression of chimeric subunits. J Biol Chem 1999; 274:27145-27152.
51. Verbitsky M, Plazas PV, Elgoyhen AB. Functional expression and properties of a nicotinic α9/5-HT3A chimeric receptor. NeuroReport 2003; 14:1931-1934.
52. Baker ER, Zwart R, Sher E et al. Pharmacological properties of α9α10 nicotinic acetylcholine receptors revealed by heterologous expression of subunit chimeras. Mol Pharmacol 2004; 65:453-460.
53. Lansdell SJ, Millar NS. Molecular characterisation of Dα6 and Dα7 nicotinic acetylcholine receptor subunits from Drosophila: formation of a high-affinity α-bungarotoxin binding site revealed by expression of subunit chimeras. J Neurochem 2004; 90:479-489.
54. Millar NS. RIC-3: a nicotinic acetylcholine receptor chaperone. Br J Pharmacol 2008; 153:S177-S183.
55. Halevi S, McKay J, Palfreyman M et al. The C. elegans ric-3 gene is required for maturation of nicotinic acetylcholine receptors. EMBO J 2002; 21:1012-1020.
56. Castillo M, Mulet J, Gutiérrez LM et al. Dual role of the RIC-3 protein in trafficking of serotonin and nicotinic acetylcholine receptors. J Biol Chem 2005; 280:27062-27068.
57. Lansdell SJ, Gee VJ, Harkness PC et al. RIC-3 enhances functional expression of multiple nicotinic acetylcholine receptor subtypes in mammalian cells. Mol Pharmacol 2005; 68:1431-1438.
58. Williams ME, Burton B, Urrutia A et al. Ric-3 promotes functional expression of the nicotinic acetylcholine receptor α7 subunit in mammalian cells. J Biol Chem 2005; 280:1257-1263.
59. Lansdell SJ, Collins T, Yabe A et al. Host-cell specific effects of the nicotinic acetylcholine receptor chaperone RIC-3 revealed by a comparison of human and Drosophila RIC-3 homologues. J Neurochem 2008; 105:1573-1581.
60. Chamaon K, Schulz R, Smalla K-H et al. Neuronal nicotinic acetylcholine receptors of Drosophila melanogaster: the α-subunit Dα3 and the β-type subunit ARD co-assemble within the same receptor complex. FEBS Lett 2000; 482:189-192.
61. Schulz R, Bertrand S, Chamaon K et al. Neuronal nicotinic acetylcholine receptors from Drosophila: two different types of α subunits coassemble within the same receptor complex. J Neurochem 2000; 74:2537-2546.
62. Chamaon K, Smalla K-H, Thomas U et al. Nicotinic acetylcholine receptors of Drosophila: three subunits encoded by genomically linked genes can co-assemble into the same receptor complex. J Neurochem 2002; 80:149-157.
63. Liu Z, Han Z, Zhang Y et al. Heteromeric co-assembly of two insect nicotinic acetylcholine receptor α subunits: influence on sensitivity to neonicotinoid insecticides. J Neurochem 2009; 108:498-506.
64. Matsuda K, Buckingham SD, Freeman JC et al. Effects of the α subunit on imidacloprid sensitivity of recombinant nicotinic acetylcholine receptors. Br J Pharmacol 1998; 123:518-524.
65. Ihara M, Matsuda K, Otake M et al. Diverse actions of neonicotinoids on chicken α7, α4β2 and Drosophila-chicken SADβ2 and ALSβ2 hybrid nicotinic acetylcholine receptors expressed in Xenopus laevis oocytes. Neuropharmacol 2003; 45:133-144.
66. Ihara M, Matsuda K, Shimomura M et al. Super agonist actions of clothianidin and related compounds on the SADβ2 nicotinic acetylcholine receptor expressed in Xenopus laevis oocytes. Biosci Biotechnol Biochem 2004; 68:761-763.
67. Raymond Delpech V, Ihara M, Coddou C et al. Action of nereistoxin on recombinant neuronal nicotinic acetylcholine receptors expressed in Xenopus laevis oocytes. Invert Neuroscience 2003; 5:29-35.
68. Shimomura M, Yokota M, Matsuda K et al. Roles of loop C and the loop B-C interval of the nicotinic receptor α subunit in its selective interactions with imidacloprid in insects. Neurosci Lett 2004; 363:195-198.
69. Shimomura M, Satoh H, Yokota M et al. Insect-vertebrate chimeric nicotinic acetylcholine receptors identify a region, loop B to the N-terminus of the Drosophila Dα2 subunit, which contributes to neonicotinoid sensitivity. Neurosci Lett 2005; 385:168-172.
70. Shimomura M, Yokota M, Ihara M et al. Role in the selectivity of neonicotinoids of insect-specific basic residues in loop D of the nicotinic acetylcholine receptor agonist binding site. Mol Pharmacol 2006; 70:1255-1263.
71. Zhang Y. Imidacloprid acts as an antagonist on insect nicotinic acetylcholine receptor containing the Y151M mutation. Neurosci Lett 2008; 446:97-100.
72. Yao X, Song F, Chen F et al. Amino acids within loops D, E and F of insect nicotinic acetylcholine receptor β subunits influence neonicotinoid selectivity. Insect Biochem Mol Biol 2008; 38:834-840.

73. Lansdell SJ, Millar NS. Dβ3, an atypical nicotinic acetylcholine receptor subunit from Drosophila: molecular cloning, heterologous expression and coassembly. J Neurochem 2002; 80:1009-1018.
74. Tomizawa M, Millar NS, Casida JE. Pharmacological profiles of recombinant and native insect nicotinic acetylcholine receptors. Insect Biochem Mol Biol 2005; 35:1347-1355.
75. Honda H, Tomizawa M, Casida JE. Insect nicotinic acetylcholine receptors: neonicotinoid binding site specificity is usually but not always conserved with varied substituents and species. J Agric Food Chem 2006; 54:3365-3371.

CHAPTER 7

Neonicotinoid Insecticides
Historical Evolution and Resistance Mechanisms

Steeve Hervé Thany*

Abstract

The use of neonicotinoid insecticides has grown considerably since their introduction in 1990s. They are used extensively for the control of agriculturally important crop pests and also in the control of cat and dog fleas. Imidacloprid exploited through an elaborated structural and substituent optimization of nithiazine was launched to market in 1990. The selectivity of neonicotinoid compounds for insect species has been attributed to their binding on nicotinic acetylcholine receptors in which the negatively charged nitro- or cyano-groups of neonicotinoid compounds interact with a cationic subsite within insect nicotinic acetylcholine receptors. The first example of a pest evolving resistance to field use of neonicotinoids was *Bemisia tabaci*. Resistance to neonicotinoids can arise either through nAChR subtypes expression, detoxification mechanisms and/or structural alterations of target-site proteins. Consequently, a number of derivatives and analogues of imidacloprid have been generated to date. In 1992, a new neonicotinoid using acetylcholine as the lead compound has been found. This was dinotefuran, which has a characteristic tetrahydro-3-furylmethyl group instead of the pyridine-like rings of others neonicotinoids.

Introduction

Pesticides encompass an array of compounds designed to prevent, destroy, repel or kill insects (insecticides), rodents (rodenticides), plants (herbicides) and fungi (fungicides). In 2004 over five billion pounds of pesticides were used worldwide. Of that amount, neonicotinoid insecticides accounted for 11-15% of the global insecticide market.[1] Neonicotinoids are the most important new class of synthetic insecticides of the past three decades, used in crop protection and animal health care due to the decrease in effectiveness of organophasphate and carbamate derivatives.[1,2] The first commercialized neonicotinoid was imidacloprid which was the largest sales of any insecticide worldwide.[2] Imidacloprid and related neonicotinoid are a promising class of insecticides with excellent biological properties, such as a wide spectrum, low application rate and quick uptake. Due to their chemical structure, they are closed to nicotine and consequently are referred to act as agonist at mammalian and insect nicotinic acetylcholine receptors (nAChRs).[3-6] Despite the fact that they are believed to be of low toxicity to mammalians because they interact much less with vertebrate nAChRs compared to insects,[7,8] ingestion of a large amount of these insecticides has been associated with the development of severe poisoning.[9-11]

*Corresponding Author: Steeve Hervé Thany—Laboratoire Récepteurs et Canaux Ioniques Membranaires (RCIM), UPRES EA 2647/USC INRA 2023, IFR 149 QUASAV, Université d'Angers, UFR de Sciences. 2 Bd. Lavoisier 49045 Angers, France.
Email: steeve.thany@univangers.fr

Insect Nicotinic Acetylcholine Receptors, edited by Steeve Hervé Thany.

Like nicotine, imidacloprid mimics the action of acetylcholine, which is the major neurotransmitter in the insect nervous system but nicotine and imidacloprid are not deactivated by acetylcholine esterase and thus persistently activate nAChRs. An analysis of insect nAChR subtypes is complicated because an unknown variety of these receptors exist and their precise subunit compositions are uncertain. Current research indicates that neonicotinoid insecticides acted as agonists, partial agonists and antagonists of insect nAChR subtypes.[4,6,12,13] In general, it has been suggested that neonicotinoid insecticides belong to the same mode of action group according to the classification developed by the Insecticide Resistance Action Committee (MOA Group 4A).[14]

Nicotine, Nicotinoids and Insecticidal Activities

Despite the considerable number of botanical insecticide reported in the literature, only four have been used for crop protection: nicotine from tobacco leaves, rotenone from derris tree roots, pyrethrum from chrysanthemum flowers and azadirachtin from neem tree. Nicotine and nicotinic-like alkaloids such as anabasine, nornicotine, piperazine alkaloids (i.e., coniine) and quinolizine alkaloids (i.e., cytisine), are present in a wide range of plant species including *Conium maculatum, Nicotiana glauca, Nicotiana tabacum, Laburnum anagyroides* and *Caulophyllum thalictroides*.[15-17] Of these plants, nicotine extracted from the tobacco plant *Nicotiana tobacum* is the most widely used, from which commercial cigarette tobacco and other nicotinic-containing devices are readily available for human consumption.[17] At the beginning of the 20th century, the significance of nicotine was increasingly studied from the toxicological, pharmacological and environmental levels. It was used for centuries as a nonsystemic insecticide to control sucking insects on plants.[18-21] Used as a foliar spray to cover the undersides of leaves, it was active as a contact and stomach poison.[21] But this botanical alkaloid was not very effective and is acutely toxic to mammals and other nontarget organisms.[21-23] In fact, nicotine and its alkaloids are readily absorbed through all routes of exposure including gastrointestinal, dermal, intranasal, inhalational routes and also through the blood brain barrier.[17,24-26] Studies on structure-activity-relationships revealed that the insecticidal activity of nicotinoids involved a 3-pyridylmethylamine moiety with a basic amino nitrogen atom as an essential structural requirement.[27,28]

In general, nicotinoids are similar to nicotine, containing an ionizable basic amine or imine substituent.[1] They include also anabaseine a marine toxin[29,30] and epibatidine isolated from the skin of an Ecuadorean frog, *Epipedobates tricolor*.[31] Others nicotinoids such as dihydronicotyrine and N,N-disubstituted 3-pyridinylmethylamine have been synthetized but not to the degree required for commercialization as insecticide.[32]

Neonicotinoid Insecticides

Screening of novel chemical structures in the 1970s and optimisation of a lead compound gave an interesting potent nitromethylene with a thiazine ring designated nithiazine (Fig. 1), the first nitromethylene insecticide, but it could not be commercialized for crop protection due to photoinstability.[23,33] However, it has served as a neonicotinoid lead structure by introducing a 6-chloropyridin-3-ylmethyl group as a substituent of the nitromethylene heterocycle which increased the insecticidal activity.[19] In fact, the discovery of imidacloprid was the result of seeking improved activity by changing the structure of 2-nitromethylenetrahydrothiazine. It became evident that the 6-chloro-3-pyridylmethyl moiety of imidacloprid attached to the saturated heterocyclic skeleton and its photostability were essential structural elements for its insecticidal activity.[34,35] The term 'neonicotinoid' was originally proposed by Izuru Yamamoto for imidacloprid and related insecticides in order to differentiate these newer insecticidally active compounds of the nAChRs from the older nicotinoid insecticides.[19,28,36]

The first major neonicotinoids were chloropyridylmethyl compounds: imidacloprid, nitenpyram, thiacloprid and acetamiprid (Fig. 2) followed soon by the nitroguanidine compound clothianidin.[1] In 1985, a series of structural modifications revealed that replacement of the

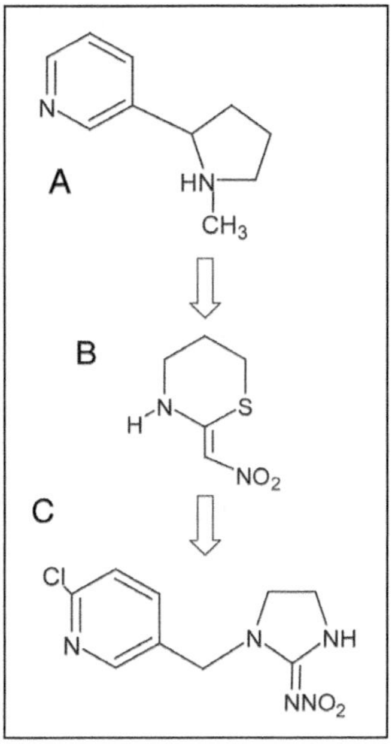

Figure 1. Development of currently used neonicotinoid insecticides from nithiazine and natural alkaloid nicotine. A) nicotine, (B) nithiazine and (C) imidacloprid.

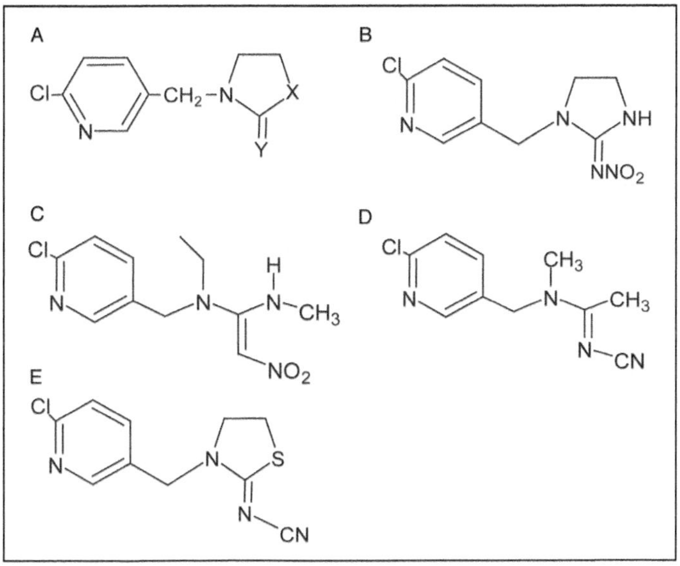

Figure 2. Chemical structures of imidacloprid (A) and related compounds (B). Y can be: NNO_2, $CHNO_2$, NCN, O or $CHCO_2Et$ and X can be: NH; NMe. Nitenpyram (C) and acetamiprid (D) have acyclic structures which differ from the imidazoline ring of imidacloprid while thiacloprid has cyclic structure (E). In addition, nitenpyram has a nitromethylene moiety in place of the nitroimino moiety of imidacloprid and acetamiprid has a cyano group in place of the electron-withdrawing nitro group of imidacloprid.

6-chloro-3-pyridyl group by a 2-chloro-5-thiazolyl moiety and the introduction of a methyl group as pharmacophore substituent resulted to the discovery of thiamethoxam, the first commercially second-generation neonicotinoid which belongs to the thianicotinyl sub-class.[37,38] Clothianidin and thiamethoxam was reported as chlorothiazolylmethyl compounds.[1,39] According to its biological features and chemical structure, it has been suggested that thiamethoxam was a neonicotinoid precursor metabolized to clothianidin in insects and plants.[40]

The distinctive structural aspect of neonicotinoids shows three important heterocyclic methyl substituents, 6-chloropyridin-3-ylmethyl, 2-chlorothiazol-5-ylmethyl and tetrahydro-3-furylmethyl which are coupled to six cyclic or acyclic N-nitroimine (=NNO_2), N-cyanoimine (=NCN) and/or 2-nitromethylene moieties (= $CHNO_2$).[1,2,41] Consequently, based on maximum inward currents induced through nAChRs activation, neonicotinoids were divided into two subgroups: (1) those with a heterocyclic ring which were partial agonists and (2) open-chain compounds which were much more effective agonists.[1,42]

Development of Novel Neonicotinoid Insecticides

Screening chemicals of novel structures in the 1970s and optimisation of a lead compound gave an interesting potent neonicotinoid reminiscent of the endogenous agonist acetylcholine (ACh). In fact, in 1990s, a new neonicotinoid using acetylcholine as the lead compound, which acts on the same receptor as nicotine, has been found. This was dinotefuran, which has a characteristic tetrahydro-3-furylmethyl group, as the hydrogen acceptor, instead of the pyridine-like rings of other neonicotinoids.[43] The binding assay and electrophysiological studies demonstrated that dinotefuran acted as agonist of insect nAChRs. In fact, binding studies of [^{3}H]epibatidine (EPI), a nAChR agonist and [^{3}H]α-bungarotoxin (α-Bgt), a competitive nAChR antagonist revealed that dinotefuran inhibited [^{3}H]EPI binding with an IC_{50} of 890 nM and [^{3}H]α-Bgt binding with an IC_{50} of 36.1 μM, indicating that dinotefuran acted as agonist of insect nAChRs.[44] This was confirmed by the finding that EPI showed a rather lower affinity to the dinotefuran binding site suggesting a high-affinity binding site in the insect.[45] Nevertheless, despite its high insecticidal activity, dinotefuran was less potent than chloropyridinyl-type or chlorothiazolyl-type neonicotinoid insecticides.[44,46] In the [^{3}H]α-Bgt assays, the difference in IC_{50} values between dinotefuran and other neonicotinoids was three orders of magnitude.[44,46,47]

A new class of insecticide targeting nAChRs but at different site from nicotine or imidacloprid has been used to increase effective control of pests.[48-51] There are spinosyns, derived from fermentation products of the actinomycete bacterium *Saccharopolyspora spinosa*.[50] Spinosad is a naturally occurring mixture of two active components, spinosyn A and spinosyn D from *S. spinosa*. Despite the fact that spinosad could act at γ-aminobutyric acid (GABA) receptors, electrophysiological studies have demonstrated that spinosad may act on nAChRs.[48] In a recent study, replacing nitromethylene pharmacophore with a conjugated system lead to series of novel neonicotinoid analogues bearing five-membered aromatic heeterocycles which exhibited higher insecticidal activities than imidacloprid.[52] Furthermore, substitutions to the phenylpyridine heterocyclic ring of nicotine lead to novel nicotinoid compounds which have insecticidal activity enhanced over that of nicotine.[53] These studies demonstrated that novel neonicotinoid analogues could be obtained, increasing the number of these active compounds.

Multiple Origins of Insect Resistance to Neonicotinoid Insecticides

The emergence of insecticide resistance mechanisms is an important question for those studying the evolution of resistance. In fact, the problem of insect resistance to insecticide was tackled by continuously introducing new active molecules to replace ones lost through resistance. But the number of biochemical sites targeted by all these compounds is limited[54] and the existence of strong resistance in some species such as *Trialeurodes vaporariorum*, the whiteflies *Bemisia tabaci*,[54-59] the Colorado potato beetle, *Leptinotarsa decemlineata*,[60-62] *Musca domestica*,[63] the western flower thrips *franklienella occidentalis*[64] and the brown planthopper *Nilaparvata lugens*[65] has demonstrated the potential of pests to adapt and resist field applications of neonicotinoids. For example, although they had never been exposed to any neonicotinoids other than imidacloprid, imidacloprid-resistant adult Colorado potato beetles, compared with a standard susceptible strain, also developed 59-fold resistance to dinotefuran, 33-fold resistance to clothianidin, 29-fold-resistance to acetamiprid, 28-fold resistance to N-methylimidacloprid, 25-fold resistance to thiacloprid and 15-fold resistance to thiamethoxam.[66] These results clearly demonstrated the existence of a cross-resistance to neonicotinoid insecticides.[61] In addition, it has been also reported that the whitefly *Bemisia tabaci* resistant strains could survive 1000-folds higher concentrations of insecticide than would be capable of killing susceptible populations.[57] They exhibited 490-fold resistance to imidacloprid and this resistance was autosomal and semi-dominant.[58] Interestingly, in the *Colorado potato* beetle, analysis of probit lines from F1 reciprocal crosses indicated that resistance to imidacloprid in adults was inherited autosomaly as an incompletely recessive factor.[60]

In general, resistance to neonicotinoids has been attributed first to mutations in nAChRs.[67-72] In the *Drosophila melanogaster* in which the entire genome was known, screening of several deficiency strains uncovering various nAChR subunit genes revealed that Df2363 strain exhibited an elevated tolerance to nitenpyram.[68] This strain was deficient across the cytological region 96A on chromosome 3 containing a cluster of *Dα1*, *Dα2* and *Dβ2* genes. Interestingly, previously it has been shown that a partial deletion of *Dα1* lacking the cytoplasmic loop, the TM4 domain and the extracellular C-terminal region fails to respond to ACh.[73] In addition, a null mutation of the drosophila Dα6 subunit confers 1181-fold resistance to Spinosad.[67] Thus, nAChR subunits were involved in neonicotinoids resistance and have critical or required elements that respond to these insecticides. This was exemplified by recent study on *Nilaparvata lugens*.[65,71] A comparison of nAChR subunits from the brown planthopper *Nilaparvata lugens* imidacloprid-sensitive and imidacloprid-resistant populations has identified a single point mutation at a conserved position Y151S in two nAChR subunits, Nlα1 and Nlα3.[71] The functional consequence of this mutation examined on recombinant Nlα1/β2, Nlα1/Nlα2/β2 and Nlα3/Nlα8/β2 receptors revealed that Y151S mutation had little or no effect on agonist potency of ACh[70,72,74] but it exerts a dramatic effect on agonist potency of neonicotinoid insecticides, when present either in Nlα1[70,72] and Nlα3.[74] The practical importance of these mutations is not clear yet, as some unidentified amino acids from other regions are involved in agonist binding.[69,75,76] Consequently it has been suggested that the X residue in loop C plays an important role in conferring high imidacloprid sensitivity, while the longer loop B-C interval region serves as a subsite supporting the nAChR-imidacloprid interactions.[75]

Unlike, the safety and effectiveness of neonicotinoids have been attributed, at least to their high affinity to nAChRs, recent studies revealed that resistance could be associated with enhanced oxidative detoxification of neonicotinoids by cytochrome P450 monooxygenases (P450).[58,60,77-80] Cytochrome P450s comprise a large superfamily of heme-thiolate proteins present in mammalians and insects that metabolize a range of both endogenous and exogenous hydrophobic compounds by incorporating oxygen into a functionalized product.[81,82] Earlier biochemical examinations revealed that neonicotinoid resistance in Q-type *Bemisia tabaci* was not associated with a lower affinity of imidacloprid to nAChRs[77,83] but to over-expression of the cytochrome P450 *CYP6CM1vQ* gene[79,80] which was capable of hydroxylating at least one position at the imidazolic ring moiety, with carbon-5, leading to 5-hydroxy form of imidacloprid.[80] Interestingly, overexpression of *CYP6g1* gene in a susceptible *D. melanogaster* strain confers imidacloprid resistance, converting it to 4-hydroxyimidacloprid and 5-hydroxyimidacloprid respectively.[84] Other enzymes in neonicotinoid biotransformations include CYP3A4 and CYP2C19, respectively.[78,85,86] CYP3A4 induces first oxidation of the imidazolidine moiety and secondarily reduction of the nitroguanidine substituent of imidacloprid.[86] Cytosolic aldehyde oxidase (AOX) which reduces the nitroguanidine moiety of imidacloprid could be considered as the neonicotinoid nitroreductase.[87] AOX system coupled with *D. melanogaster* nAChR inactivates clothianidin, dinotefuran or imidacloprid.[78] Consequently, the coupled nicotinic receptor-metabolic system could be a predictor for toxicity and/or detoxifying mechanisms in insects. In the other hand, resistance against neonicotinoids insecticides could be conferred by changes in detoxicative enzymes.

Conclusion

In recent years, new compounds have been identified. Nereistoxin analogues (cartap and monosultap) and spynosyns (spinosad) are commercial insecticide classes targeting insect nAChRs. In fact, Spinosad, a naturally occurring mixture of two macrocyclic lactones (spinosyn A and spinosyn D) isolated from the soil actinomycete *Saccharopolyspora spinosa*, has been developed as a neonicotinoid insecticide against lepidopteran and thysanopteran pests. It provided a inherent broad spectrum against several insect including fruit flies, tobacco budworm and *Harmonia axyridis* (harlequin ladybird).[48,49] In conclusion, because the subunit combinations and the tissue distributions of insect nAChR subtypes are poorly understood, our understanding of how resistance to neonicotinoid insecticides might occur through nAChR activation was uncertain. In fact, neonicotinoid actions and efficacies were strongly correlated to the way they were applied, the concentrations used and the intracellular mechanisms modulating nAChR activation.

References

1. Tomizawa M, Casida JE. Neonicotinoid insecticide toxicology: Mechanisms of selective action. Annu Rev Pharmacol Toxicol 2005; 45:247-68.
2. Tomizawa M, Casida JE. Selective toxicity of neonicotinoids attributable to specificity of insect and mammalian nicotinic receptors. Annu Rev Entomol 2003; 48:339-64.
3. Bunckingham SD, B. L, Le Corronc H et al. Imidacloprid actions on insect neuronal acetylcholine receptors. J Exp Biol 1997; 200:2685-91.
4. Ihara M, Matsuda K, Otake M et al. Diverse actions of neonicotinoids on chicken alpha7, alpha4beta2 and drosophila-chicken sadbeta2 and alsbeta2 hybrid nicotinic acetylcholine receptors expressed in Xenopus laevis oocytes. Neuropharmacology 2003; 45(1):133-44.
5. Ihara M, Matsuda K, Shimomura M et al. Super agonist actions of clothianidin and related compounds on the SAD beta2 nicotinic acetylcholine receptor expressed in xenopus laevis oocytes. Biosci Biotechnol Biochem 2004; 68(3):761-3.
6. Thany SH. Agonist actions of clothianidin on synaptic and extrasynaptic nicotinic acetylcholine receptors expressed on cockroach sixth abdominal ganglion. Neurotoxicology 2009 (In press).
7. Lansdell SJ, Millar NS. The influence of nicotinic receptor subunit composition upon agonist, alpha-bungarotoxin and insecticide (imidacloprid) binding affinity. Neuropharmacology 2000; 39(4):671-9.
8. Tomizawa M, Cowan A, Casida JE. Analgesic and toxic effects of neonicotinoid insecticides in mice. Toxicol Appl Pharmacol 2001; 177(1):77-83.
9. Wu IW, Lin JL, Cheng ET. Acute poisoning with the neonicotinoid insecticide imidacloprid in n-methyl pyrrolidone. J Toxicol Clin Toxicol 2001; 39(6):617-21.
10. Shadnia S, Moghaddam HH. Fatal intoxication with imidacloprid insecticide. Am J Emerg Med 2008; 26(5):634 e1-4.
11. Phua DH, Lin CC, Wu ML et al. Neonicotinoid insecticides: An emerging cause of acute pesticide poisoning. Clin Toxicol 2009; 47(4):336-41.
12. Deglise P, Grunewald B, Gauthier M. The insecticide imidacloprid is a partial agonist of the nicotinic receptor of honeybee Kenyon cells. Neurosci Lett 2002; 321(1-2):13-6.
13. Zhang Y, Liu S, Gu J et al. Imidacloprid acts as an antagonist on insect nicotinic acetylcholine receptor containing the Y151M mutation. Neurosci Lett 2008; 446(2-3):97-100.
14. Classification IMoA. Insecticide resistance action committee (IRAC). 2005.
15. Kisaki T, Tamaki E. Phytochemical studies on the tobacco alkaloids. I. Optical rotatory power of nornicotine. Arch Biochem Biophys 1961; 92:351-355.

16. Lopez TA, Cid MS, Bianchini ML. Biochemistry of hemlock (conium maculatum l.) alkaloids and their acute and chronic toxicity in livestock. A review. Toxicon 1999; 37(6):841-65.
17. Schep LJ, Slaughter RJ, Beasley DM. Nicotinic plant poisoning. Clin Toxicol 2009; 47(8):771-81.
18. Yamamoto I. Nicotinoids as insecticides. In: Metcalf RL, ed. In Advances in Pest Control Research, Vol. 6. New York: Wiley, 1965:231-260.
19. Yamamoto I, Tomizawa M, Saito T et al. Structural factors contributing to insecticidal and selective actions of neonicotinoids. Arch Insect Biochem Physiol 1998; 37(1):24-32.
20. Casida JE, Quistad GB. Golden age of insecticide research: Past, present, or future? Annu Rev Entomol 1998; 43:1-16.
21. Leonard GC, Julius JM. Biopesticides: A review of their action, applications and efficacy. Pest Management Science 2000; 56(8):651-676.
22. Schmeltz I. Nicotine and other tobacco alkaloids. In: Jacobson M, Crosby DG, eds. Naturally Occurring Insecticides. New York: Dekker, M, 1971:99-136.
23. Soloway SB. Naturally occurring insecticides. Environ Health Perspect 1976; 14:109-17.
24. Luck W, Nau H. Exposure of the fetus, neonate and nursed infant to nicotine and cotinine from maternal smoking. N Engl J Med 1984; 311(10):672.
25. Luck W, Nau H. Nicotine and cotinine concentrations in serum and milk of nursing smokers. Br J Clin Pharmacol 1984; 18(1):9-15.
26. Reavill C, Walther B, Stolerman IP et al. Behavioral and pharmaco-kinetic studies on nicotine, cytisine and lobeline. Neuropharmacology 1990; 29:619-624.
27. Tomizawa M, Yamamoto I. Binding of nicotinoids and the related compounds to the insect nicotinic acetylcholine receptors. J Pestic Sci 1992; 17:231-236.
28. Tomizawa M, Yamamoto I. Structure activity relationships of neonicotinoids and imidacloprid analogues. J Pestic Sci 1993; 18:91-98.
29. Kem WR, Mahnir VM, Papke RL et al. Anabaseineis a potent agonist upon muscle and neuronal alpha-bungarotoxin sensitive nicotinic receptors. J Pharmacol Exp Ther 1997; 283:979-992.
30. Kem WR, Mahnir VM, Prokai L et al. Hydroxy metabolites of the Alzheimer's drug candidate 3-[(2,4-dimethoxy)benzylidene]-anabaseine dihydrochloride (GTS-21): Their molecular properties, interactions with brain nicotinic receptors and brain penetration. Mol Pharmacol 2004; 65(1):56-67.
31. Spande TF, Garraffo HM, Edwards MW et al. Epibatidine: A novel (chloropyridyl)azabicycloheptane with potent analgesic activity from an ecuadorean poison frog. J Am Chem Soc 1992; 114:3475-3478.
32. Tomizawa M, Zhang N, Durkin KA et al. The neonicotinoid electronegative pharmacophore plays the crucial role in the high affinity and selectivity for the drosophila nicotinic receptor: An anomaly for the nicotinoid cation—pi interaction model. Biochemistry 2003; 42(25):7819-27.
33. Kleier D, Holden I, Casida JE et al. Novel photoreactions of an insecticidal nitromethylene heterocycle. J Agric Food Chem 1985; 33:998-1000.
34. Moriya K, Shibuya K, Hattori Y et al. Structural modification of the 6-chloropyridyl moiety in the imidacloprid skeleto: Introduction of a five-membered heteroaromatic ring and the resulting insecticidal activity. Biosci Biotechnol Biochem 1993; 57(1):127-128.
35. Kagabu S, Medej S. Stability comparison of imidacloprid and related compounds under stimulated sunlight, hydrolysis conditions and to oxygen. Biosci Biotechnol Biochem 1995; 59:980-985.
36. Yamamoto I. Nicotinoids as insecticides. Adv Pest Control Res 1965; 6:231-260.
37. Maienfisch P, Huerlimann H, Rindlisbacher A et al. The discovery of thiamethoxam: A second-generation neonicotinoid. Pest Manag Sci 2001; 57(2):165-76.
38. Maienfisch P, Angst M, Brandl F et al. Chemistry and biology of thiamethoxam: A second generation neonicotinoid. Pest Manag Sci 2001; 57(10):906-13.
39. Ford KA, Casida JE. Unique and common metabolites of thiamethoxam, clothianidin and dinotefuran in mice. Chem Res Toxicol 2006; 19(11):1549-56.
40. Nauen RE-K U, Salgado VL, Kaussmann M. Thiamethoxam is a neonicotinoid precursor converted to clothianidin in insects and plants. Pestic Biochem Physiol 2003; 76:55-69.
41. Honda H, Tomizawa M, Casida JE. Insect nicotinic acetylcholine receptors: Neonicotinoid binding site specificity is usually but not always conserved with varied substituents and species. J Agric Food Chem 2006; 54(9):3365-71.
42. Tan J, Galligan JJ, Hollingworth RM. Agonist actions of neonicotinoids on nicotinic acetylcholine rececptors expressed by cockroach neurons. NeuroToxicology 2007; 28:829-842.
43. Wakita T, Yasui N, Yamada E et al. Development of a novel i nsecticide, dinotefuran. J Pestic Sci 2005; 30(2):112-123.

44. Mori K, Okumoto T, Kawahara N et al. Interaction of dinotefuran and its analogues with nicotinic acetylcholine receptors of cockroach nerve cords. Pest Manag Sci 2002; 58(2):190-6.
45. Miyagi S, Komaki I, Ozoe Y. Identification of a high-affinity binding site for dinotefuran in the nerve cord of the american cockroach. Pest Manag Sci 2006; 62(4):293-8.
46. Zhang A, Kayser H, Maienfisch P et al. Insect nicotinic acetylcholine receptor: Conserved neonicotinoid specificity of [(3)h]imidacloprid binding site. J Neurochem 2000; 75(3):1294-303.
47. Kiriyama K, Nishiwaki H, Nakagawa Y et al. Insecticidal activity and nicotinic acetylcholine receptor binding of dinotefuran and its analogues in the housefly, Musca domestica. Pest Manag Sci 2003; 59(10):1093-100.
48. Salgado VL, Sheets JJ, Watson GB et al. Studies on the mode of action of spinosad: The internal effective concentration and concentration dependence of neural excitation. Pest Biochem Physiol 1998; 60:103-110.
49. Salgado VL. Studies on the mode of action of spinosad: Insect symptoms and physiological correlates. Pest Biochem Physiol 1998; 60:91-102.
50. Thompson GD, Dutton R, Sparks TC. Spinosad a case study: An example from a natural products discovery program. Pest Manag Sci 2000; 56:696-702.
51. Galvan TL, Koch RL, Hutchison WD. Toxicity of indoxacarb and spinosad to the multicolored asian lady beetle, harmonia axyridis (coleoptera: Coccinellidae), via three routes of exposure. Pest Manag Sci 2006; 62(9):797-804.
52. Shao X, Li Z, Qian X et al. Design, synthesis and insecticidal activities of novel analogues of neonicotinoids: Replacement of nitromethylene with nitroconjugated system. J Agric Food Chem 2009; 57(3):951-7.
53. Stumpf CF, Comins DL, Sparks TC et al. Insecticidal activity and mode of action of novel nicotinoids synthesized by new acylpyridiniumsalt chemistry and directed lithiation. Pest Biochem Physiol 2007; 87:211-219.
54. Nauen R, Denholm I. Resistance of insect pests to neonicotinoid insecticides: Current status and future prospects. Arch Insect Biochem Physiol 2005; 58(4):200-15.
55. Elbert A, Nauen R. Resistance of Bemisia tabaci (homoptera: Aleyrodidae) to insecticides in southern spain with special reference to neonicotinoids. Pest Manag Sci 2000; 60(1):60-64.
56. Gorman K, Devine G, Bennison J et al. Report of resistance to the neonicotinoid insecticide imidacloprid in Trialeurodes vaporariorum (hemiptera: Aleyrodidae). Pest Manag Sci 2007; 63(6):555-8.
57. Roditakis E, Grispou M, Morou E et al. Current status of insecticide resistance in q biotype Bemisia tabaci populations from crete. Pest Manag Sci 2009; 65(3):313-22.
58. Wang Z, Yao M, Wu Y. Cross-resistance, inheritance and biochemical mechanisms of imidacloprid resistance in b-biotype Bemisia tabaci. Pest Manag Sci 2009.
59. Schuster DJ, Mann RS, Toapanta M et al. Monitoring neonicotinoid resistance in biotype b of Bemisia tabaci in florida. Pest Manag Sci 2009.
60. Zhao JZ, Bishop BA, Grafius EJ. Inheritance and synergism of resistance to imidacloprid in the colorado potato beetle (coleoptera: Chrysomelidae). J Econ Entomol 2000; 93(5):1508-14.
61. Alyokhin A, Dively G, Patterson M et al. Resistance and cross-resistance to imidacloprid and thiamethoxam in the Colorado potato beetle Leptinotarsa decemlineata. Pest Manag Sci 2007; 63(1):32-41.
62. Tan J, Salgado VL, Hollingworth RM. Neural actions of imidacloprid and their involvement in resistance in the colorado potato beetle, Leptinotarsa decemlineata (say). Pest Manag Sci 2008; 64(1):37-47.
63. Kristensen M, Jespersen JB. Susceptibility to thiamethoxam of Musca domestica from danish livestock farms. Pest Manag Sci 2008; 64(2):126-32.
64. Zhao G, Liu WEI, Brown JM et al. Insecticide resistance in field and laboratory strains of western flower thrips (thysanoptera: Thripidae). J Econ Entomol 1995; 88:1164-1170.
65. Zewen L, Zhaojun H, Yinchang W et al. Selection for imidacloprid resistance in Nilaparvata lugens: Cross-resistance patterns and possible mechanisms. Pest Manag Sci 2003; 59(12):1355-9.
66. Mota-Sanchez D, Hollingworth RM, Grafius EJ et al. Resistance and cross-resistance to neonicotinoid insecticides and spinosad in the colorado potato beetle, Leptinotarsa decemlineata (say) (coleoptera: Chrysomelidae). Pest Manag Sci 2006; 62(1):30-7.
67. Perry T, McKenzie JA, Batterham P. A dalpha6 knockout strain of Drosophila melanogaster confers a high level of resistance to spinosad. Insect Biochem Mol Biol 2007; 37(2):184-8.
68. Perry T, Heckel DG, McKenzie JA et al. Mutations in dalpha1 or dbeta2 nicotinic acetylcholine receptor subunits can confer resistance to neonicotinoids in Drosophila melanogaster. Insect Biochem Mol Biol 2008; 38(5):520-8.

69. Liu Z, Han Z, Liu S et al. Amino acids outside of the loops that define the agonist binding site are important for ligand binding to insect nicotinic acetylcholine receptors. J Neurochem 2008; 106(1):224-30.
70. Liu Z, Han Z, Zhang Y et al. Heteromeric co-assembly of two insect nicotinic acetylcholine receptor alpha subunits: Influence on sensitivity to neonicotinoid insecticides. J Neurochem 2009; 108(2):498-506.
71. Liu Z, Williamson MS, Lansdell SJ et al. A nicotinic acetylcholine receptor mutation conferring target-site resistance to imidacloprid in Nilaparvata lugens (brown planthopper). Proc Natl Acad Sci USA 2005.
72. Liu Z, Williamson MS, Lansdell SJ et al. A nicotinic acetylcholine receptor mutation (y151s) causes reduced agonist potency to a range of neonicotinoid insecticides. J Neurochem 2006; 99(4):1273-81.
73. Schulz R, Bertrand S, Chamaon K et al. Neuronal nicotinic acetylcholine receptors from drosophila: Two different types of alpha subunits coassemble within the same receptor complex. J Neurochem 2000; 74(6):2537-46.
74. Yixi Z, Liu Z, Han Z et al. Functional co-expression of two insect nicotinic receptor subunits (Nlalpha3 and Nlalpha8) reveals the effects of a resistance-associated mutation (Nlalpha3) on neonicotinoid insecticides. J Neurochem 2009; 110(6):1855-62.
75. Shimomura M, Yokota M, Matsuda K et al. Roles of loop C and the loop B-C interval of the nicotinic receptor alpha subunit in its selective interactions with imidacloprid in insects. Neurosci Lett 2004; 363(3):195-8.
76. Shimomura M, Satoh H, Yokota M et al. Insect-vertebrate chimeric nicotinic acetylcholine receptors identify a region, loop B to the N-terminus of the drosophila dalpha2 subunit, which contributes to neonicotinoid sensitivity. Neurosci Lett 2005.
77. Rauch N, Nauen R. Identification of biochemical markers linked to neonicotinoid cross resistance in Bemisia tabaci (hemiptera: Aleyrodidae). Arch Insect Biochem Physiol 2003; 54(4):165-76.
78. Honda H, Tomizawa M, Casida JE. Neonicotinoid metabolic activation and inactivation established with coupled nicotinic receptor-Cyp3a4 and -aldehyde oxidase systems. Toxicology Letters 2006; 161(2):108-114.
79. Karunker I, Benting J, Lueke B et al. Over-expression of cytochrome P450 Cyp6cm1 is associated with high resistance to imidacloprid in the B and Q biotypes of Bemisia tabaci (hemiptera: Aleyrodidae). Insect Biochem Mol Biol 2008; 38(6):634-44.
80. Karunker I, Morou E, Nikou D et al. Structural model and functional characterization of the Bemisia tabaci Cyp6cm1vq, a cytochrome P450 associated with high levels of imidacloprid resistance. Insect Biochem Mol Biol 2009.
81. Tijet N, Helvig C, Feyereisen R. The cytochrome P-450 gene superfamily in Drosophila melanogaster: Annotation, intron-exon organization and phylogeny. Gene 2001; 262:189-198.
82. Berenbaum MR. Postgenomic chemical ecology: From genetic code to ecological interactions. J Chem Ecol 2002; 28(5):873-96.
83. Nauen R, Stumpf N, Elbert A. Toxicological and mechanistic studies on neonicotinoid cross resistance in Q-type Bemisia tabaci (hemiptera: Aleyrodidae). Pest Manag Sci 2002; 58(9):868-75.
84. Joussen N, Heckel DG, Haas M et al. Metabolism of imidacloprid and DDT by P450 Cyp6g1 expressed in cell cultures of nicotiana tabacum suggests detoxification of these insecticides in Cyp6g1-overexpressing strains of Drosophila melanogaster, leading to resistance. Pest Manag Sci 2008; 64(1):65-73.
85. Schulz-Jander DA, Leimkuehler WM, Casida JE. Neonicotinoid insecticides: Reduction and cleavage of imidacloprid nitroimine substituent by liver microsomal and cytosolic enzymes. Chem Res Toxicol 2002; 15(9):1158-65.
86. Schulz-Jander DA, Casida JE. Imidacloprid insecticide metabolism: Human cytochrome P450 isozymes differ in selectivity for imidazolidine oxidation versus nitroimine reduction. Toxicol Lett 2002; 132(1):65-70.
87. Dick RA, Kanne DB, Casida JE. Identification of aldehyde oxidase as the neonicotinoid nitroreductase. Chem Res Toxicol 2005; 18(2):317-23.

Chapter 8

Ecotoxicity of Neonicotinoid Insecticides to Bees

Axel Decourtye* and James Devillers

Abstract

This chapter reviews the available data on the toxicity of neonicotinoid insecticides to bees that are the prominent and the most economically important group of pollinators worldwide. Classical and new methods developed to take into account the characteristics and different types of effects of the neonicotinoid insecticides to bees are described. The available toxicity results are critically analyzed. Thus, the nitro-substituted compounds (clothianidin, dinotefuran, imidacloprid and its metabolites, thiamethoxam, nitenpyram) appear the most toxic to bees. The cyano-substituted neonicotinoids seem to exhibit a much lower toxicity (acetamiprid and thiacloprid). The chapter ends with suggestions for additional studies aiming at better assess the hazard of this important insecticide family to bees.

Introduction

Bees are the predominant and the most economically important group of pollinators worldwide. Thirty five per cent of world crop production depend on pollinators.[1] In Europe, the production of 84% of crop species depends at least to some extent upon animal pollination.[2] Bees serve humanity indirectly by contributing to the healthy functioning of unmanaged terrestrial ecosystems. The decline of pollinating species, which have grown over the last decades,[3] can lead to a parallel decrease of plant species.[4] More specifically, there is a great deal of concern about the decline of the honey bee (*Apis mellifera* L.) across the world that has been termed colony collapse disorder.[5] The abundance of pollinators in the environment is influenced by biotic factors (predators, pathogens, parasites, competitors, availability of resources) and abiotic factors (climate, pollutants). Although the putative causes of this decline are still currently analyzed,[5] it is admitted that the extensive use of pesticides against pest insects for crop protection has contributed to the loss of many pollinators.[6] Pollinating insects, such as the honey bee, are mainly exposed to chemicals when visiting melliferous plants. Consequently, for several decades, following the development and use of pesticides, honey bees have focused interest in the frame of the protection of nontargeted organisms against pesticide damages and the first guidelines dealing with this aspect were published in the 1950s.[7] Nowadays, the hazard assessment of pesticide toxicity to honey bees is commonly estimated from laboratory studies (median lethal dose: LD50) and from semi-field and field experimentations when the pesticides demonstrate a hazard quotient (application rate/LD50) greater than 50, or when they show a specific mode of action (e.g., insect growth regulators), or when there are indications

*Corresponding Author: Axel Decourtye—ACTA, Institut Claude Bougelat, 1 Avenue C. Bourgelat, 68280 Marcy L'Etoile, France. Email: axel.decourtye@acta.asso.fr

Insect Nicotinic Acetylcholine Receptors, edited by Steeve Hervé Thany.

of indirect effects such as delayed action.[8] Despite these preregistration studies, neonicotinoid insecticides were recently implicated by beekeepers who reported that hives placed near cropped plants, originated from seeds dressed with insecticide, showed high levels of damage due to a progressive decrease in the hive populations, until the complete loss of the colonies.[9] The risk that systemic neonicotinoid insecticides induce for honey bees started in France with the use of Gaucho® (active ingredient: imidacloprid) on sunflower.[10] Imidacloprid was the first insecticide belonging to the new chemical family of neonicotinoids and has become a widely used chemical. Next, the controversy was spread over other neonicotinoid insecticides, other pollinators and the decline of bees worldwide. Thus, in recent years, numerous studies have been performed to assess whether seed dressings containing neonicotinoid insecticides could be harmful to honey bees.

This chapter reviews the available data on the toxicity of neonicotinoid insecticides to bees. The interest of the new ecotoxicological methods developed in order to take into account the characteristics of neonicotinoid insecticides is also highlighted.

Effects on Survival

Acute Toxicity

The standard method to evaluate the toxicity of the insecticides that can potentially be in contact with the honey bees consists in the calculation of an acute toxicity data (LD50) following standardized guidelines (European Council Directive 91/414 in Europe and the Federal Insecticide Fungicide and Rodenticide Act in the United States). In this context, laboratory bioassays were conducted to determine the oral and contact honey bee toxicity of neonicotinoid insecticides (Table 1). The LD50 values of neonicotinoids are low compared with older families of insecticides such as organophosphorus, pyrethroids, carbamates.[11,12] They are considered highly toxic to honey bees.[13] Generally, neonicotinoids are more toxic via oral route than contact mode. The difference between the oral and contact toxicity may be due to the weak hydrophobicity of the neonicotinoids yielding a low penetration through the insect cuticle. The contact LD50 values of imidacloprid for the honey bee do not vary significantly from an author to another (a factor of 6 of variation). Using the same application route, Stark et al showed that three species of bees, *Apis mellifera, Megachile rotundata and Nomia melanderi*, were equally susceptible to imidacloprid (24-h LD50 = 0.04 μg/bee).[11] Similar results were obtained for Admire and Provado that are two commercial formulations of imidacloprid.[14,15] On the other hand, the oral LD50 value of imidacloprid may vary widely (factor 20) in the honey bee. Since the oral toxicity test is based on the homogeneous repartition of the chemical in the group of bees by trophallaxis, the neurotoxicity symptoms observed for some of them, which could be due to the action of imidacloprid and not of its metabolites,[16] might result in the ingestion of unequal doses of toxic. This could explain the high variability in the LD50 values sometimes recorded between tests. According to Nauen et al,[17] this fact can also explain the delayed toxicity at higher doses of imidacloprid that was observed by Suchail et al. In order to test the hypothesis of the lack of trophallaxis process, the oral LD50 of imidacloprid was determined from a collective versus individual treatment (Decourtye, unpublished data).[12] The 48-h oral LD50 values of imidacloprid were similar after collective (25.4 ± 22.8 ng per bee) and individual treatment (25.1 ± 22.8 ng per bee). It appears that the heterogeneous distribution of toxic in bee group was not a decisive factor for the determination of imidacloprid toxicity. This underlines the assumption that other factors can bias the LD50. For example, the effects of imidacloprid vary according to the age of bees,[17,18] the subspecies[12] or colony.[19] All these factors are not well standardized in the routine toxicity tests.

Table 1. Acute oral and contact toxicity (48h) of neonicotinoid insecticides to honey bees

Insecticide	Oral LD50 (μg/Bee)	Contact LD50 (μg/Bee)
Acetamiprid	14.53 (European commission)*	8.09 (European commission) 7.07[20]
Clothianidin	0.003 (European commission)	0.044 (European commission) 0.022[20]
Dinotefuran	Unknown	0.023 (Footprint)** 0.075[20]
Imidacloprid	0.004-0.041[21] >0.08[17] 0.0057[19] 0.005[19] 0.003[22]	0.030[23] 0.018[20] 0.081-0.23 Two ascending parts of the dose-effect curve: 0.007 and 0.024[12] 0.043-0.104[17]
Olefin (imidacloprid metabolite)	0.028[19] 0.003[22]	0.04[11]
5-OH-imidacloprid (imidacloprid metabolite)	0.258[19] 0.153[22]	
Nitenpyram	Unknown	0.138[20]
Thiacloprid	17.32 (European commission)	38.82 (European commission) 14.6[20]
Thiamethoxam	0.005 (European commission)	0.024 (European commission) 0.03[20]

*http://e-phy.agriculture.gouv.fr/ **http://sitem.herts.ac.uk/aeru/footprint/fr/

Chronic Toxicity

The acute toxicity tests would only account for a situation where foragers are exposed to high dose/short term treatments. But in the case of systemic compounds, such as neonicotinoids, longer term effects are not excluded, since the product is potentially present in the nectar or pollen of plants seed dressing with these insecticides.[21,24,25] Hive worker bees may also be exposed to the chemicals since foragers collect potentially contaminated food to be stored inside the hive. As the stored food originates from different plants, a dilution of toxic compounds occurs, however they can be present in the hive at lower concentration but for longer periods than on plants. Thus, acute toxicity is an incomplete measure of the adverse effects of systemic insecticides because of the short duration of these tests that are generally of 1 to 3 days. Therefore, many studies have examined the long-term effects of neonicotinoid insecticides, with an ecologically relevant exposure, on survival of honey bees in order to develop assays mimicking realistic conditions. Thus, in a study performed by Aliouane et al, experiments were conducted with bees collected from hives, caged in groups of 30-60 individuals and provided with sucrose solution contaminated or not during 10-11 consecutive days.[26] The mortality was recorded daily. Chronic oral exposure to thiamethoxam or acetamiprid had no effect on mortality. On the other hand, chronic oral intoxication experiments have shown that imidacloprid and six metabolites (olefin, 5-hydroxy-imidacloprid, 4,5-dihydroxy-imidacloprid, 6-chloronicotinic acid, desnitroimidacloprid, urea derivative) induced mortality at concentrations of 0.1, 1 and 10 $\mu g\, l^{-1}$.[19] These results are not in agreement with those obtained by Schmuck et al who reported that any lethal effect was recorded above 20 μg l^{-1} of imidacloprid and its plant metabolites.[27] This high discrepancy in the results may be due to differences in the experimental methodologies such as the use of bees with specified age or not, a randomization procedure that can be applied or not,[27] or the large variability in the toxic effects according to the physiological state of individuals exposed to the chemical.[16] An additional cause deals with the fact that bees do not die independently of each other. Food exchanges, contacts and pheromonal communication occurring among workers induce that the survival duration of a bee may depend on the survival duration of its nestmates.[28] This influence of social interactions was integrated in a Cox proportional hazard model designed by Dechaume Moncharmont and coworkers.[29] With 60-day dietary exposure, a treatment effect was found for bees exposed to 4 and 8 μg l^{-1} of imidacloprid. This work showed that variability between replicates and density-dependent effect could not be neglected in the assessment of chronic toxicity of pesticides to the honey bee.

Larvae Toxicity

A lot of studies focus on the effects of neonicotinoids to adult bees but reductions in brood may be more damaging to colony health than the loss of old bees, such as foragers. Indeed, flexibility in the division of labor can replace foragers if there are enough brood and nurse bees. Because a gap in brood rearing may, therefore, cause the decline of the colony, a method for evaluating the side-effects of plant protection products on honey bee larvae is recommended in the official guidelines, especially for compounds having insect growth-regulating properties.[8] Even if no insect growth-regulating property was attributed to neonicotinoid insecticides, some works have reported effects on larvae development. The developmental time necessary for the emergence of larvae fed with imidacloprid-added food (5 μg kg^{-1}) was longer than in controls.[30] This delay in the development of larvae exposed to imidacloprid (30-300 μg kg^{-1}) was confirmed with *Osmia lignaria*.[31] Moreover, Taséi et al noted a reduction of the number of larvae produced in colonies of *Bombus terrestris* fed with syrup and pollen containing imidacloprid at 10 and 6 μg kg^{-1}, respectively.[32]

Behavioral Effects

The behavioral effects of neonicotinoid insecticides were largely investigated over the last ten years. This fact is linked to the depopulations of hives that have been observed by beekeepers near fields sowed with seed-dressing treated plants. It was assumed that foragers collecting nectar and pollen were exposed to low doses of neonicotinoid insecticides during their foraging trips, which induced behavioral effects and subsequently no homing return to hive. So, many studies have been carried out in order to assess the effects of pesticides on behavioral traits and more particularly on the foraging behavior of bees treated with the insecticides accused by beekeepers.

Mobility

Many symptoms reveal a general mobility affected by a neurotoxic molecule: knockdown, uncoordinated movements (or staggering), trembling, tumbling, abdomen tucking, rotating and cleaning of abdomen while rubbing hind legs together, decreased walking. A publication reporting acute toxicity tests of imidacloprid in the honey bee carried out by seven different European research facilities indicated knockdown effects. Moreover, a lot of bees were immobile and hence, unable to feed.[17] These symptoms of intoxication were observed at higher doses and persisted for a maximum of 48 h after the treatment. These observations were consistent with other early symptoms of poisoning that appeared after oral ingestion of imidacloprid, such as stationary or inactive behavior,[33,34] movement coordination problems, tumbling, hyperactivity and tremors.[19] After several hours, hyperactivity gradually disappears and the bees become hypoactive.[19] However, imidacloprid can also induce opposite effects on motor activity depending on the dose.[35] In laboratory, these authors recorded the position of worker bee in an open-field-like allowing observation of bee vertical displacements. Inverse effects of imidacloprid on motor activity were observed depending on the dose. The lowest dose (1.25 ng per bee) induced an increase of motor activity whereas the higher doses (2.5 to 20 ng per bee) induced a decrease of displacements in the box. Among the nitroguanidine neonicotinoid group, thiamethoxam did not induce this activating effect after acute[36] or chronic exposure.[26] Under similar experimental conditions, El Hassani et al found that acetamiprid increased locomotion activity (0.1 and 0.5 μg/bee).[36] But this effect was not confirmed by Aliouane et al.[26] The test of open-field is based on negative geotaxis or positive photoaxis since honey bees tend to migrate upward against the force of gravity to the light source. This test is relevant to assess motor function of walking bees, but definitively not suited to evaluate the flying activity, which is a process of the foraging behavior suspected to be altered by neonicotinoids.

Learning Performances

When landing on a flower, the forager extends its proboscis as a reflex when the gustatory receptors set on the tarsae, antennae or mouth-parts are stimulated with nectar. This reflex leads to the uptake of nectar and induces the memorization of the floral odors diffusing concomitantly. Thus, the memorization of odors plays a prominent role in flower recognition during the next trips.[37] The olfactory learning, involved in flower recognition, can be studied in laboratory with a bioassay based on the conditioning of the proboscis extension reflex (PER) applied to restrained individuals.[38] The PER assay with restrained workers has been used to investigate the behavioral effects of about 20 pesticides[39] including neonicotinoid insecticides.[40-42] An acute exposure of bees can be applied before, during or after the PER conditioning.[42] But, the long-term exposure is more suited to the characteristics of neonicotinoid compounds. It corresponds to the case of bees that are newly involved in foraging duties based on their learning ability, after being fed a contaminated food within the hive. With this approach, reduced learning performances were observed for bees surviving to 11 days of oral treatment with imidacloprid, 5-OH-imidacloprid,[22] acetamiprid,[36] and of topic treatment with thiamethoxam.[26] With the PER assay, we can also investigate how a chemical treatment can interfere on the memory process. It has been shown that acetamiprid induced long-term memory impairment after an oral absorption.[36] Chronic contact with thiamethoxam (0.1 μg per bee) induced a decrease of memory 24 h after learning, followed by a recovery at 48 h that rules out long-term memory impairment.[26] Decourtye et al showed that imidacloprid administered acutely impaired the medium-term olfactory memory.[41] The structure-specific increase of cytochrome oxidase activity into the brain that is observed after treatment with imidacloprid suggests an impairment of olfactory memory provoked by a physiological effect at the higher cerebral level, in the area called "mushroom bodies".[41,43] But, the precise consequences of these results for the foraging behavior are still unclear. There is a problem in how the results can be extrapolated to what should happen in realistic field situation. In general, results from these studies cannot be extrapolated to natural conditions. Moreover, imidacloprid can also have facilitatory effects on learning performances that complicate the interpretation at an ecological level. Indeed, the habituation procedure has been used to demonstrate the facilitatory effect of a sublethal dose of imidacloprid on the PER suppression.[35,44] Habituation of the PER is a simple form of learning in which the repetition of the gustatory stimulation leads to a decrease of the response probability. These results clearly indicate task dependent behavioral effects of sublethal concentrations of imidacloprid.

Orientation

Honey bees can navigate accurately and repeatedly to a food source, as well as communicate to their nestmates the distance and direction in which to fly to reach it.[45] The process of foraging involves learning and memory, communication, navigation, taking into account information from the internal clock and many other flexible responses such as the ability to integrate local landmarks.[46] A bee exposed to pesticide during foraging trip can have a wrong acquisition or integration of these neurobiological processes. This is particularly true for the visual learning of landmarks, which is important in spatial orientation. One of the major tasks for the honey bee during a foraging flight is to learn and recall many complex visual patterns.[47] It is well known that honey bees use landmark-based cues to navigate to a goal and to return to the nest. These cues are needed to set the flight direction, to monitor progress to the goal, to provide intermediate guiding landmarks and finally aid in spatial tracking the target when the bee is in its vicinity.[48] Considering these neurobiological functions in orientation processes, it is of great interest to know whether neurotoxic neonicotinoids induce behavioral disturbances and if these alterations exist at low concentration level. To test whether thiamethoxam may disorientate foragers, its impact was examined on orientation of honey bees in a maze under outdoor conditions. Orientation performance of bees in a complex maze relies on associative learning between a visual mark and a reward of sugar solution.[49] Bees had to fly through a sequence of nine boxes to reach a goal—a feeder containing a reward of sugar solution. Along the path, boxes constituted a decision point: the bee had to choose between a marked hole (correct path) and unmarked hole leading to dead end. Correct or wrong decisions and turns back were recorded in foragers orally with 3 ng per bee of thiamethoxam (Decourtye, unpublished data). While, 61% of no-treated bees equally flied through the whole path and arrived to the goal without mistakes, this rate decreased to 38% in thiamethoxam-treated bees. Conversely, the rate of bees with unsuccessful searches for goal and with turns back, increased with treatment. Our results show that orientation capacities of foragers in a complex maze are affected by thiamethoxam. But thiamethoxam is not a direct-acting agonist or antagonist of nAchR[50] and clothianidin, its toxic metabolite, causes the biological effect.[20,51] Unlike in the maze where the performances are based on the use of limited pertinent cues, the navigation in the field relies on several guidance mechanisms. Bees are able to recognize patterns in situations where local landmarks are not reliable.[52] Additional experiments are needed to establish whether foragers exposed to neonicotinoid insecticides can negotiate a route in a complex environment or if they are lost, this being a possible cause in the drastic bee population losses as observed by beekeepers. For that, field studies have recorded the displacements of foragers trained to forage on an artificial feeder filled with a sucrose solution. Bees foraging on the syrup were captured and individually marked with colored tags that were numbered. After release, the homing flight of the bees was tracked. With a distance between feeder and hive of 500 m, foragers fed with 500 $\mu g\ kg^{-1}$ and 1000 $\mu g\ kg^{-1}$ of imidacloprid were seen neither at the hive nor at the feeding site, for the 24 hours after the treatment.[53] Foragers fed with imidacloprid-added syrup at the concentration of 100 $\mu g\ kg^{-1}$ had a delay for returning to their hive or feeding site for up to 24 h. A more recent behavioral study has confirmed these results but with a shorter distance. Imidacloprid with concentration above 50 $\mu g\ l^{-1}$ affected the interval between two successive visits of the same bee to the feeding site.[54] Imidacloprid added in sucrose solution with concentration above 600 $\mu g\ l^{-1}$ caused failure in return to the feeding site of foragers. In these field studies, it is likely that the foragers exposed to concentration of imidacloprid as low as 500 $\mu g\ kg^{-1}$ got lost and died somewhere in the field, but the lowest observed effect concentration on the frequentation of feeding site was 50 $\mu g\ kg^{-1}$. Concerning the exposure to imidacloprid and its metabolites, the values were 3.3 and 3.5 $\mu g\ kg^{-1}$ in the pollen of Gaucho®-treated sunflowers and maize, respectively and 1.9 $\mu g\ kg^{-1}$ in the nectar of Gaucho®-treated sunflowers.[55]

Foraging and Feeding Behaviors

It is known that bees might change their behavior in response to their sensory perception of pesticides, by reduced foraging or a feeding stimulation.[56] In most cases repellency induced by pesticides is considered as a protective behavior effective in reducing the risk associated with these potentially dangerous chemicals. However, special attention must be paid to avoid that the repellency of pesticides affects the economical value of *Apis* bees and non-*Apis* bees as crop pollinating agents. For example, the harmless of neonicotinoid treatments of tomatoes was verified on the pollinating rate provided by *Bombus terrestris* in a greenhouse with imidacloprid[57] and with thiamethoxam.[58] *Bombus impatiens* workers exposed to imidacloprid showed no effect at field residue levels.[59] Observations on foraging activity of *Bombus terrestris* revealed no difference regarding the presence of workers on blooming heads and the duration of their visits between imidacloprid-treated and control sunflowers.[32] Another member of the chloronicotinyl family, clothianidin appeared harmless when colonies of *B. impatiens* were exposed to the highest residue levels in pollen,[60] or when hives of *Apis mellifera* were exposed to flowering canola grown from clothianidin-treated seeds.[61] Although these studies showed the absence of effect of neonicotinoids on foraging of treated plants, perturbations of the foraging behavior on artificial feeder were revealed in other experiments. Thus, for example, it was found a quick decrease in the foraging activity in honey bee colonies at about 20 ppb of imidacloprid.[40,62] This is probably due to the anti-feedant character of the compound. Several authors have confirmed that foraging bees reduced their visits to a syrup feeder when it was contaminated with concentrations above 20 ppb.[63,64] But, the delay in the inhibition of foraging with imidacloprid varies according to the concentration tested.[62] The author suggested that the delay was due to the process occurring inside the hive rather than effects on the foragers. This hypothesis was reinforced by a study reporting a reduction in the foraging activity on a food source contaminated with imidacloprid (20-100 ppb) due to the induction of trembling dances that prevent other bees from foraging and the decrease in the frequency of wagging dances, which provide then the recruitment of foragers for a food sources.[64] Thus, the changes in the communication process can result in a decreased foraging activity.

The motor and sensory functions of the honey bee are linked to the foraging capacities of bees. The forager must be able to perceive the sugar of the nectar then extends its tongue (called "proboscis") to take nectar. Therefore, induced modifications of one of these functions by pesticide exposure may have repercussions on foragers' activity leading to a disruption of nectar collect. For example, laboratory studies reported the effects of chronic intoxication of young honey bees with sublethal doses of acetamiprid and thiamethoxam on sensory perception of sugar.[26] Responsiveness to antennal sucrose stimulation was decreased for high sucrose concentrations in honey bees treated orally with thiamethoxam (1 ng per bee), but not with acetamiprid.

Conclusion

Many agricultural crops and natural plant populations depend on pollination and often on the services provided by pollinator communities.[65-67] The causes of decline among pollinators vary from a species to another and are generally difficult to assign. Definitive causes of decline can be assigned in only a few cases.[5] Declines of many pollinators are associated with diseases, pollution, habitat deterioration and combination of these different causes. One factor contributing to pollinator declines is pesticides[5,6] and this context, neonicotinoid-treated crops could be hazardous to managed and wild bees that feed on pollen and nectar-containing residues. Systemic neonicotinoid compounds differ from the classical sprayed insecticides, which are present on the plant only several hours or days after application. With systemic neonicotinoids, the exposure of honey bee to chemicals is possible during several weeks of flowering. Especially, the effects of repeated consumption of contaminated stocks of nectar or pollen inside the hive can appear either immediately or with delayed effects. The impact of Gaucho®-treated sunflowers in the honey bee in France initiated a large series of scientific studies.[10] Next, the question was extended to other uses of neonicotinoid insecticides, other pollinators and the decline of bees across the world. To date, a number of studies have recorded measurements of short or long term toxicity, lethal or sublethal effects of these insecticides to bees. In summary, the nitro-substituted compounds (clothianidin, dinotefuran, imidacloprid and its metabolites, thiamethoxam, nitenpyram) appear the most toxic to the honey bee. The cyano-substituted neonicotinoids seem to exhibit a much lower toxicity (acetamiprid and thiacloprid). Amongst the neonicotinoids, imidacloprid has been the most studied. A high discrepancy in results of chronic feeding studies was reported with imidacloprid and its metabolites. Suchail et al related the similarity of the observed dietary toxicity between metabolites with and without the binding site on nAchR to the chloropyridine structure, supporting the hypothesis of a novel pharmacological mechanism unrelated to this binding site.[19] Schmuck has contradicted this fact based on repetitions of chronic study of Suchail et al cited above and on receptor binding and electrophysiologic studies.[27]

On the other hand, the exposure of bees due to the presence of neonicotinoid residues in the pollen or nectar of treated plants would induce possible behavioral effects during foraging trips. As a result, an increasing number of studies and methods related to the identification and characterization of sublethal effects of neonicotinoids have been published in the past 10 years. This chapter has revealed new insights into behavioral effects of neonicotinoids including effects on mobility, orientation, foraging and learning. The lowest observed effect concentrations for imidacloprid reported were the following: 3 $\mu g\ kg^{-1}$ for foraging activity[34] and 12 $\mu g\ kg^{-1}$ for olfactory learning.[22] But these data are often inadequate to demonstrate causation unambiguously. It is not clear whether the endpoints tested in these sublethal studies can be clearly related to the respective field effect of concern.[39,68] Conversely, for the methods based on orientation and homing ability and that were proposed to be tested on bees, the ecological relevance is better.[53,54] If measurable differences in homing flight are found in such experimental design, an individual, exposed under field conditions would probably suffer from this effect. For instance, foragers fed with imidacloprid-added syrup at the concentration of 100 $\mu g\ kg^{-1}$ showed a delay for returning to their hive.[53] Such acute exposure cannot probably occur in the realistic conditions since the concentrations of imidacloprid and its metabolites, to which honey bees are exposed always have been measured lower than 10 $\mu g\ kg^{-1}$.[21,24,25,62,69,71,72] Although data in field tests are often inadequate to demonstrate causation unambiguously, the loss of the colonies could not be verified for imidacloprid under field conditions, where no effects on hive development have been detected.[9,73,74] The effect of neonicotinoid insecticides on bee health still remains highly controversial, especially imidacloprid.[10] In the light of growing evidence that pollinators are exposed to pollen and nectar-containing residues in field conditions, the contribution of neonicotinoid insecticides to the decline of pollinators warrants additional investigations.

Acknowledgements

This work was supported in part by a grant from the European Community program (1221/97) for French beekeeping co-ordinated by French Ministry of Agriculture.

References

1. Klein AM, Vaissière B, Cane JH et al. Importance of crop pollinators in changing landscapes for world crops. Proc R Soc Lond B Biol Sci 2007; 274:303-313.
2. Williams IH. The dependence of crop production within the European Union on pollination by honey bees. Agric Zool Rev 1994; 6:229-257.
3. Goulson D, Lye GC, Darvill B. Decline and conservation of bumble bees. Annu Rev Entomol 2008; 53:11.1-11.18.

4. Biesmeijer JC, Roberts SPM, Reemer M et al. Parallel declines in pollinators and insect-pollinated plants in Britain and the Netherlands. Science 2006; 313:351-354.
5. Committee on the Status of Pollinators in North America. Status of Pollinators in North America 2006.
6. Allen-Wardell G, Bernhardt P, Bitner R et al. The potential consequences of pollinator declines on the conservation of biodiversity and stability of food crop yields. Conserv Biol 1998; 12:8-17.
7. Cluzeau S. Risk assessment of plant protection products on honey bees. In: Devillers J, Pham-Delègue MH, eds. Honey Bees: Estimating the Environmental Impact of Chemicals. London: Taylor and Francis 2002:42-55.
8. European and Mediterranean Plant Protection Organism. Guideline on test methods for evaluating the side-effects of plant protection products on honey bees. EPPO Bull 1992; 22:203-215.
9. Maus C, Curé G, Schmuck R. Safety of imidacloprid seed dressings to honey bees: a comprehensive overview and compilation of the current state of knowledge. Bull Insect 2003; 56(1):51-57.
10. Maxima L, van der Sluijs JP. Uncertainty: Cause or effect of stakeholders' debates?: Analysis of a case study: The risk for honeybees of the insecticide Gaucho®. Sci Total Environ 2007; 376(1-3):1-17.
11. Stark JD, Jepson PC, Mayer DF. Limitation to use of topical toxicity data for prediction of pesticide side effect in the field. J Econ Entomol 1995; 88:1081-1088.
12. Suchail S, Guez D, Belzunces LP. Characteristics of imidacloprid toxicity in two Apis mellifera subspecies. Environ Toxicol Chem 2000; 19:1901-1905.
13. Atkins EL, Kellum D, Atkins KW. Reducing pesticide hazards to honey bees: mortality prediction techniques and integrated management strategies. Division of Agricultural Sciences University of California USA 1981; 3-22.
14. Mayer DF, Johansen CA. How to reduce bee poisoning from pesticides, Pacific Northwest Extension Publication, PNW 1999; 518:14.
15. Devillers J, Decourtye A, Budzinski H et al. Comparative toxicity and hazards of pesticides to Apis and non-Apis bees. A chemometrical study, SAR QSAR Environ Res 2003; 14:389-403.
16. Suchail S, Debrauwer L, Belzunces LP. Metabolism of imidacloprid in Apis mellifera. Pest Manag Sci 2003; 60:291-296.
17. Nauen R, Ebbinghaus-Kintscher U, Elbert A et al. Acetylcholine receptors as sites for developing neonicotinoid insecticides. In: Ishaaya I ed. Biochemical sites important in insecticide action and resistance, Springer Verlag Berlin 2001; 77-105.
18. Guez D, Belzunces LP, Maleszka R. Effects of imidacloprid metabolites on habituation in honeybees suggest the existence of two subtypes of nicotinic receptors differentially expressed during adult development. Pharmacol Biochem Behav 2003; 75:217-222.
19. Suchail S, Guez D, Belzunces LP. Discrepancy between acute and chronic toxicity induced by imidacloprid and its metabolites in Apis mellifera. Environ Toxicol Chem 2001; 20:2482-2486.
20. Iwasa T, Motoyama N, Ambrose JT et al. Mechanism for the differential toxicity of neonicotinoid insecticides in the honey bee, Apis mellifera. Crop Protect 2004; 23:371-378.
21. Schmuck R, Schöning R, Stork A et al. Risk posed to honeybees (Apis mellifera L, Hymenoptera) by an imidacloprid seed dressing of sunflowers. Pest Manag Sci 2001; 57:225-238.
22. Decourtye A, Lacassie E, Pham-Delègue MH. Learning performances of honeybee (Apis mellifera L) are differentially affected by imidacloprid according to the season. Pest Manag Sci 2003; 59:269-278.
23. Ruzhong G, Rui C, Liangyan C. Evaluation on toxicity and safety of imidacloprid to environmental organisms. Pestic Sci Administration 1999; 20(3).
24. Wallner K. Test regarding effects of imidacloprid on honey bees. In: Belzunces LP, Pélissier C, Lewis G, eds. Proceedings of the 7th International Symposium "Hazards of pesticides to bees", Les Colloques de l'INRA 2001; 98:91-94.
25. Bonmatin JM, Moineau I, Charvet R et al. A LC/APCI-MS/MS method for analysis of imidacloprid in soil, in plants and in pollens. Anal Chem 2003; 75:2027-2033.
26. Aliouane Y, El Hassani AK, Gary V et al. Subchronic exposure of honeybees to sublethal doses of pesticides: effects on behaviour. Environ Toxicol Chem 2009; 28(1):113-122.
27. Schmuck R. Effects of a chronic dietary exposure of the honeybee Apis mellifera (Hymenoptera: Apidae) to imidacloprid. Arch Environ Contam Toxicol 2004; 47:471-478.
28. Wilson EO. The insect societies. Cambridge: Harvard University Press 1971.
29. Dechaume-Moncharmont FX, Decourtye A, Hennequet C et al. Statistical analysis of the honeybee survival after chronic exposure to insecticides. Environ Toxicol Chem 2003; 22(12):3088-3094.

30. Decourtye A, Tisseur M, Taséi JN et al. Toxicité et risques liés à l'emploi de pesticides chez les pollinisateurs : cas de l'abeille domestique. In: Regnault-Roger C. ed. Enjeux phytosanitaires pour l'agriculture et l'environnement au XXI° siècle. Tec et Doc Lavoisier, Paris 2005; 283-299.
31. Abbott VA, Nadeau JL, Higo HA et al. Lethal and sublethal effects of imidacloprid on Osmia lignaria and clothianidin on Megachile rotundata (Hymenoptera: Megachilidae). J Econ Entomol 2008; 101(3):784-796.
32. Taséi JN, Ripault G, Rivault E. Hazards of imidacloprid seed coating to Bombus terrestris (Hymenoptera: Apidea) when applied to sunflower. J Econ Entomol 2001; 94:623-627.
33. Medrzycki P, Montanari R, Bortolotti P et al. Effects of imidacloprid administered in sub-lethal doses on honey bee behaviour. Laboratory tests. Bull Insect 2003; 56(1):59-62.
34. Colin ME, Bonmatin JM, Moineau I et al. A Method to quantify and analyze the foraging activity of honey bees: relevance to the sublethal effects induced by systemic insecticides. Arch Environ Contam Toxicol 2004; 47:387-395.
35. Lambin M, Armengaud C, Raymond S et al. Imidacloprid-induced facilitation of the proboscis extension reflex habituation in the honeybee. Arch Insect Biochem Physiol 2001; 4:129-134.
36. El Hassani AK, Dacher M, Gary V et al. Effects of sublethal doses of acetamiprid and thiamethoxam on the behavior of the honeybee (Apis mellifera). Arch Environ Contam Toxicol 2008; 54:653-661.
37. Menzel R, Greggers U, Hammer M. Functional organization of appetitive learning and memory in a generalist pollinator, the honey bee. In: Papaj DR, Lewis AC eds. Insect learning. New York: Chapman Hall 1993: 79-125.
38. Takeda K. Classical conditioned response in the honey bee. J Insect Physiol 1961; 6:168-179.
39. Desneux N, Decourtye A, Delpuech JM. The sublethal effects of pesticides on beneficial arthropods. Annu Rev Entomol 2007; 52:81-106.
40. Decourtye A, Devillers J, Cluzeau S et al. Effects of imidacloprid and deltamethrin on associative learning in honeybees under semi-field and laboratory conditions. Ecotoxicol Environ Saf 2004; 57(3):410-419.
41. Decourtye A, Armengaud C, Renou M et al. Imidacloprid impairs memory and brain metabolism in the honey bee (Apis mellifera L.). Pestic Biochem Physiol 2004; 78(2):83-92.
42. Decourtye A, Pham-Delègue MH. The proboscis extension response: assessing the sublethal effects of pesticides on the honey bee. In: Devillers J, Pham-Delègue MH eds. Honey bees: estimating the environmental impact of chemicals, Taylor and Francis publ., London 2002; 67-84.
43. Armengaud C, Causse N, Aït-Oubah J et al. Functional cytochrome oxidase histochemistry in the honeybee brain. Brain Research 2000; 859:390-393.
44. Guez D, Suchail S, Maleszka R et al. Contrasting effects of imidacloprid on habituation in 7-day and 8-day old honeybees. Neurobiol Learn Memory 2001; 76:183-191.
45. von Frisch K. The dance language and orientation of bees. Cambridge:Harvard University Press 1967.
46. Menzel R, Geiger K, Joerges J et al. Bees travel novel homeward routes by integrating separately acquired vector memories. Anim Behav 1998; 55:139-152.
47. Capaldi EA, Dyer FC. The role of orientation flights on homing performance in honeybees. J Exp Biol 1999; 202:1655-1666.
48. Collett TS, Fry SN, Wehner R. Sequence learning by honey bees. J Comp Physiol A 1993; 172, 693-706.
49. Zhang SW, Bartsch K, Srinivasan MV. Maze learning by honeybees. Neurobiol Learn Mem 1996; 66:267-282.
50. Tan J, Galligan JJ, Hollingworth RM. Agonist actions of neonicotinoids on nicotinic acetylcholine receptors expressed by cockroach neurons. Neurotoxicol 2007; 28:829-842.
51. Nauen R, Ebbinghaus-Kintscher U, Salgado V et al. Thiamethoxam is neonicotinoid precursor converted to clothianidin in insects and plants. Pest Biochem Physiol 2003; 76:55-69.
52. Lindauer M. Time-compensated sun orientation in bees. Cold Spring Harbor Symp quant Bio 1960; 25:371-377.
53. Bortolotti L, Montanari R, Marcelino J et al. Effect of sub-lethal imidacloprid doses on the homing rate and foraging activity of honey bees. Bull Insect 2003; 56:63-67.
54. Yang EC, Chuang YC, Chen YL et al. Abnormal foraging behavior induced by sublethal dosage of imidacloprid in the honey bee (Hymenoptera: Apidae). J Econ Entomol 2008; 101(6):1743-1748.
55. Rortais A, Arnold G, Halm MP et al. Modes of honeybees exposure to systemic insecticides: estimated amounts of contaminated pollen and nectar consumed by different categories of bees. Apidologie 2005; 36:71-83.

56. Haynes KF. Sublethal effects of neurotoxic insecticides on insect behavior. Annu Rev Entomol 1988; 33:149-68.
57. Colombo A, Buonocore E. Effetto di trattamenti al terreno con imidacloprid sull'attività dei bombi. L'Informatore Agrario 1997; 53(38):85-87.
58. Alarcón AL, Cánovas M, Senn R et al. The safety of thiamethoxam to pollinating bumble bees (Bombus terrestris L.) when applied to tomato plants through drip irrigation. Commun Agric Appl Biol Sci 2005; 70(4):569-79.
59. Morandin LA, Winston ML. Effects of novel pesticides on bumble bee (Hymenoptera: Apidae) colony health and foraging ability. Environ Entomol 2003; 32:555-563.
60. Franklin MT, Winston ML, Morandin LA. Effects of clothianidin on Bombus impatiens (Hymenoptera: Apidae) colony health and foraging ability. J Econ Entomol 2004; 97(2):369-73.
61. Cutler GC, Scott-Dupree CD. Exposure to clothianidin seed-treated canola has no long-term impact on honey bees. J Econ Entomol 2007; 100(3):765-72.
62. Schmuck R. No causal relationship between Gaucho® seed dressing in sunflowers and the French bee malady. Pflanzenschutz-Nachrichten Bayer 1999; 52:267-309.
63. Mayer DF, Lunden JD. Effects of imidacloprid insecticide on three bee pollinators. Hortic Sci 1997; 29:93-97.
64. Kirchner WH. Mad-bee-disease? Sublethal effects of Imidacloprid ("Gaucho") on the behaviour of honey-bees. Apidologie 1999; 30:422.
65. Díaz S, Tilman D, Fargione J et al. Biodiversity regulation of ecosystem services. In: Hassan R, Scholes R, Ash N, eds. Ecosystems and human well-being: Current state and trends: Findings of the Condition and Trends Working Group Washington DC: Island Press 2005; 297-329.
66. Biesmeijer JC, Roberts SPM, Reemer M et al. Parallel declines in pollinators and insect-pollinated plants in Britain and the Netherlands. Science 2006; 313:351-354.
67. Gallai N, Salles JM, Settele J et al. Economic valuation of the vulnerability of world agriculture confronted with pollinator decline. Ecol Econ 2009; 68:810-821.
68. Thompson HM, Maus C. The relevance of sublethal effects in honey bee testing for pesticide risk assessment. Pest Manag Sci 2007; 63:1058-1061.
69. Curé G, Schmidt HW, Schmuck R. Results of a comprehensive field research programme with the systemic insecticide imidacloprid (Gaucho). In: Belzunces LP, Pelissier C, Lewis GB, eds. Proceedings of the 7th International Symposium "Hazards of pesticides to bees", Les Colloques de l'INRA 2001; 98:49-59.
70. Bonmatin JM, Marchand PA, Charvet R et al. Quantification of imidacloprid uptake in maize crops. J Agric Food Chem 2005; 53:5336-5341.
71. Laurent FM, Rathahao E. Distribution of [14C]imidacloprid in sunflower (Helianthus annuus) following seed treatment. J Agric Food Chem 2003; 51: 8005-8010.
72. Stadler T, Martinez Ginés D, Buteler M. Long-term toxicity assessment of imidacloprid to evaluate side effects on honey bees exposed to treated sunflower in Argentina, Bull Insect 2003; 56:77-81.
73. Faucon JP, Aurières C, Drajnudel P et al. Experimental study on the toxicity of imidacloprid given in syrup to honey bee (Apis mellifera) colonies, Pest Manag Sci 2005; 61:111-125.

CHAPTER 9

State of the Art on Insect Nicotinic Acetylcholine Receptor Function in Learning and Memory

Monique Gauthier*

Abstract

Acetylcholine is the most abundant excitatory neurotransmitter in the insect brain and the most numerous acetylcholine receptors are the nicotinic ones (nAChRs). The genome sequencing of diverse insect species has demonstrated the existence of at least 10 nAChR genes coding for α and β subunits, suggesting the existence in the insect CNS of several subtypes of nAChRs whose molecular composition and pharmacological properties are still unknown. Insect nAChRs have given rise to an abundance of literature about their sensitivity to neonicotinoid insecticides but only limited data are available on the functional role of nAChRs in insect cognitive functions. The data we have collected on honeybees are the only data that shed light on the role of nAChRs in learning and memory processes. The behavioral response of proboscis extension (PER), which appears when the honeybee perceives sugar, was used to quantify learning and memory performances in associative and non-associative learning procedures. Habituation of the PER, which consists in ceasing to respond to sucrose upon repetitive antennal sucrose stimulation, was facilitated by the injection into the brain of one of the nicotinic antagonists mecamylamine, alpha-bungarotoxin (α-BGT) or methyllycaconitine (MLA). Pavlovian associative protocol was used to condition the PER to olfactory or tactile stimulus after single- or multiple-trial training. Localized brain injections of the nicotinic antagonist mecamylamine were performed before or after one-trial olfactory learning in the mushroom bodies (MB), the integrative structures of the insect brain. The results showed that the calical input structures of the MB are necessary for the acquisition processes and the output α-lobe regions are involved in retrieval processes. Brain injection of one of the three nicotinic antagonists mecamylamine, α-BGT and MLA was combined with single- and multiple-trial olfactory and tactile learning and memory performances were evaluated at long- or short-term intervals. Mecamylamine impaired the acquisition of one-trial learning and the retrieval of information, regardless of the number of trials during training and the learning modality (olfactory or tactile cues used as conditioned stimulus). Memory performance evaluated at long-term intervals was decreased by injection of α-BGT and MLA in multiple-trial olfactory and tactile experiments. We conclude from these results that at least two subtypes of nAChRs exist in the honeybee brain. The α-BGT-sensitive nAChRs are necessary for the formation of long-term memory and the α-BGT-insensitive nAChRs are involved in one-trial acquisition

*Monique Gauthier—Centre de Recherches sur la Cognition Animale (CRCA), UMR 5169 CNRS UPS, 118 route de Narbonne 31062 Toulouse cedex, France. Email: gauthiem@cict.fr

Insect Nicotinic Acetylcholine Receptors, edited by Steeve Hervé Thany.

and in retrieval processes. The hypothesis is put forward that multiple-trial associative learning triggers activation of the α-BGT-sensitive nAChRs that, in turn, activate intracellular events leading to LTM formation.

Introduction

In vertebrates the literature devoted to the role of acetylcholine (ACh) in brain functions and diseases is impressive and well documented.[1-4] Acetylcholine binds to nicotinic and muscarinic acetylcholine receptors widely distributed in the central nervous system (CNS). There are fewer nicotinic acetylcholine receptors (nAChRs) than muscarinic ones.[4] The nAChRs are formed from the assembly of five subunits arranged in such a way as to delimit a central cationic channel. As nine α (α2-10) and three β (β2-4) subunits are identified in the CNS, various combinations of subunits will produce many different nAChR subtypes with different pharmacological and biophysical properties.[4] The neuronal nAChRs can be broadly separated into two forms, the heteromeric receptors formed from the association of α and β subunits and the homomeric receptors constituted from the assembly of five identical α subunits. In vertebrates, the most strongly represented heteromeric receptor is the α4β2 subtype, which is sensitive to the non-competitive open-channel blocker mecamylamine and to di-hydro-β-erythroidine. The most strongly represented homomeric receptor is the α7 subtype, which is specifically blocked by the snake venom α-bungarotoxin (α-BGT) and by methyllycaconitine (MLA). In the CNS, the nAChRs are mainly presynaptic and modulate the release of a number of neurotransmitters. They are also found postsynaptically and they control fast ACh-mediated synaptic transmission.[5] The α4β2 receptors are largely distributed throughout the brain and they are mainly associated with pain control, nicotine addiction and Parkinson disease. Alpha 7 subunits and resulting subtypes are more concentrated in the forebrain, frontal cortex and hippocampus and are involved in schizophrenia, epilepsy and anxiety. Both types are implicated in memory processes and are proposed as potential therapeutic targets for the treatment of Alzheimer's disease.[4]

In insects, ACh is the most abundant neurotransmitter in the CNS.[6] As opposed to glutamate in vertebrates, ACh is thought to be the primary excitatory neurotransmitter in the CNS. This is attested in honeybees,[7] flies,[8] grasshoppers[9] and locusts.[10] Insect nervous tissue is one of the richest sources of neuronal nAChRs.[11] These receptors outnumber the muscarinic receptors in the insect brain and are widely distributed in the synaptic neuropile regions. Immunocytochemical and binding studies have shown that major sensory and integrative structures of the bee brain are cholinergic[12-17] (Fig. 1). Staining for acetylcholinesterase (AChE), the enzyme that hydrolizes ACh, has been found in the antennal lobe (AL) glomeruli, where primary olfactory afferents from the antennae make synaptic contact with local neurons and projection neurons that send olfactory information to mushroom bodies (MB), the integrative structures of the protocerebrum. The median antennoglomerular tract formed from the axons of the projection neurons that connect the glomeruli with the lip area of the median and lateral calyx of the MB is stained. The dorsal lobes that house antennal mechanosensory and gustatory[18] terminals are also stained. Nicotinic AChR-immunoreactivity (nAChR-IR) is found in MB, specifically in the lip region of the calyces that receive olfactory projections from the antennal lobes. The MB intrinsic neurons, the Kenyon cells, send output axons through the pedunculus, forming the α and the β lobes. Alpha lobes show nAChR-IR. The output fibers from the pedunculus connect to the lateral protocerebrum and then to the suboesophageal ganglion, which houses the motoneurons that command the mouthpart movements. The suboesophageal ganglion is weakly stained for nAChR-IR.

Electrophysiological recordings of ACh-induced currents from insect neurons, in culture or in situ, have been performed in a variety of insect species (for review see refs. 19, 20). In the honeybee the currents are induced by the classical nicotinic agonists (nicotine, epibatidine) and equally blocked by the antagonists mecamylamine, hexamethonium, α-BGT and MLA.[21-25]

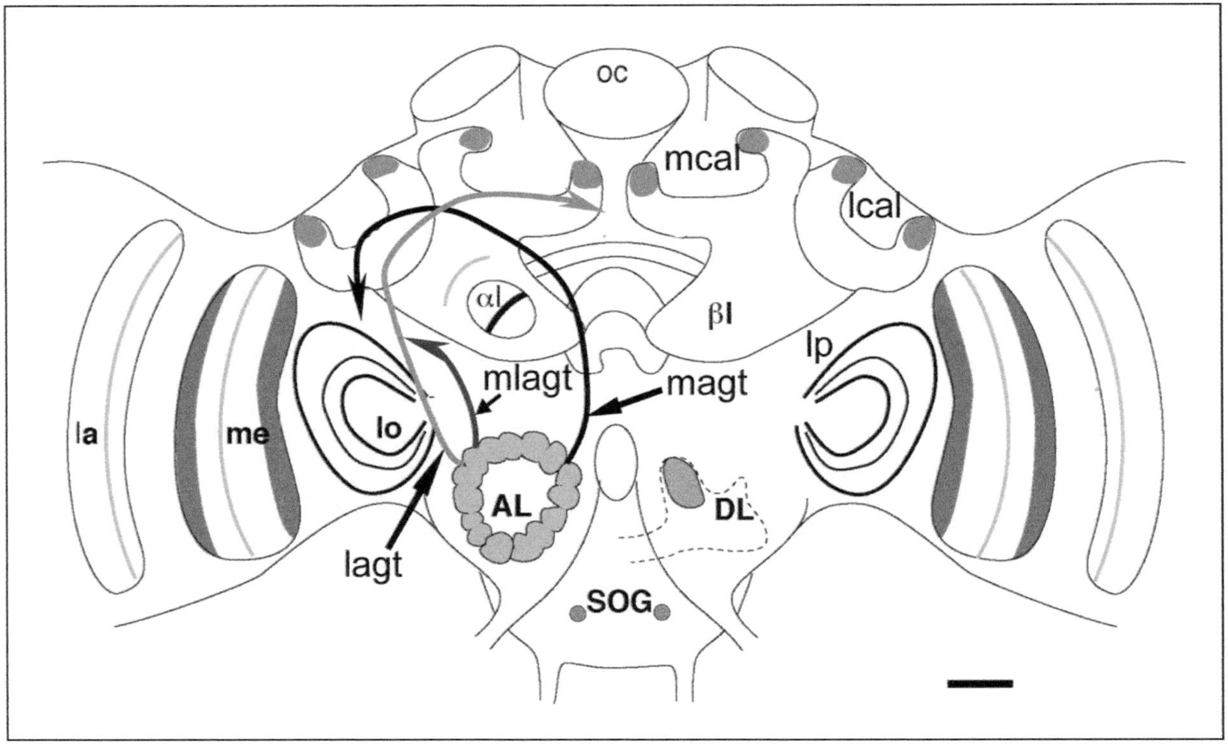

Figure 1. Schematic drawing of labeling for AChE activity and ACh receptor-like immunoreactivity in sensory and integrative structures of the bee brain in a frontal view. Stained areas appear in grey. Optic lobes: lamina (la), medulla (me), lobula (lo). Antennal lobe (AL): glomeruli are stained. Mushroom bodies: the lip region of the median calyce (mcal) and the lateral calyce (lcal) is stained; alpha lobe (αl) is stained; beta lobe (βl). Three tracts connect the ALs to the calyces of mushroom bodies and to lateral protocerebrum (lp): the median antennoglomerular tract (magt), which is stained, the lateral antennoglomerular tract (lagt) and the mediolateral antennoglomerular tract (mlagt). Dorsal lobe (DL). Suboesophageal ganglia (SOG). Median ocella (OC). Scale: 100 μm.

Analyses of genome sequences have shown that nAChR gene families remain compact in diverse insect species.[26] *Drosophila melanogaster* and *Anopheles gambiae* each possess 10 nAChR genes while *Apis mellifera* has 11 genes, 9 α and 2 β.[27] These subunits can co-assemble in different combinations to form several subtypes of nAChRs with distinct pharmacological properties. The molecular composition and the stoechiometry of the insect nAChRs are unknown and the pharmacology of in vitro native receptors does not help in differentiating nAChR subtypes. Insect nAChRs have given rise to an abundance of literature about their sensitivity to insecticides[28,29] but only a limited amount of data is available concerning the functional roles of nAChRs in insect behavioral and cognitive functions. This chapter will review the literature concerning the involvement of nAChRs in insect behavior and in learning and memory processes. For this latter part, I will report data collected from our work on the honeybee, which is the only research being done in the field.

Role of ACh and Nicotinic Acetylcholine Receptors in Insect Behavior

Sensory and motor functions depend upon activation of central cholinergic pathways in insects. Cholinergic transmission via central synapses has been associated with mechanosensory reflexes in the locust. Pressure ejection of acetylcholine or its agonists carbachol and nicotine into the prothoracic neuropile mimicked wind stimulation and elicited excitation of a ventral cord interneuron that makes output synapses to motor neurons of special wing steering muscle. The fast response is mediated in part by nAChRs[10] and is similar to fast excitatory responses observed in many other insects. In grasshoppers, nicotinic activation of neuropile regions of the protocerebrum and the suboesophageal ganglia produces wing movements responsible for stridulation, with rapid onset no different from a natural song.[9,30] Application to honeybees of insecticides belonging to the neonicotinoid class such as imidacloprid, acetamiprid or thiamethoxam has shed light on the role of the cholinergic system in sensory and motor functions. Acute low doses of imidacloprid or acetamiprid increased locomotor displacements.[31-33] Acetamiprid also enhanced water and sugar responsiveness, whereas sub-lethal doses of thiamethoxam had no effect on these functions.[33] These results suggest that activation of the nicotinic system with low doses of neonicotinoid may enhance behavioral functions. They also indicate that the three compounds present different affinities to the honeybee nAChRs. Recently, cholinergic local neurons have been identified within the *Drosophila* antennal lobe glomeruli. The hypothesis has been put forward that these neurons contribute to olfactory coding by broadening the odor tuning of projection neurons compared to the odor tuning of their presynaptic olfactory receptor neurons.[34,35] The cholinergic system has also been associated with addictive processes in insects. The fruit fly *Drosophila melanogaster* is presented as an attractive model system to investigate the mechanisms of addiction. In this respect, studies have been carried out on the effects of nicotine on startle-induced climbing behavior, which is the flies' ability to negatively geotax in a column when the column is gently knocked on a soft surface to force the flies to the bottom. The reflexive climbing response is impaired by nicotine exposure,[36] an effect amplified by repetitive exposures to nicotine.[37] Genetic and pharmacological evidence suggests that the enhancement of the nicotinic effect is dependent on a mechanism involving the cAMP/PKA cascade.[37]

Role of Nicotinic Acetylcholine Receptors in Learning and Memory in the Honeybee

Very little research is devoted to the study of ACh and its role in insect behavioral plasticity supported by learning and memory. All the studies that do exist use the honeybee as an experimental model (this chapter and refs. 38-42); this is due to the rich behavioral repertoire exhibited by the honeybee in natural or laboratory conditions.[43] Moreover, the honeybee offers the possibility of being trained individually, which is an ideal condition for developing studies on learning and measuring the effect of this individual experience on behavior. We made wide use of the proboscis extension reflex

(PER) to study the role of the nAChRs in memory processes. In natural conditions, the proboscis extension is a natural response that occurs when gustatory receptors of antennae, mouthparts or tarsus are touched with a feeding sucrose solution. The obvious consequence for the honeybee is the possibility of consuming food. In unusual conditions like those prevailing in laboratory experiments, the PER is elicited in restrained bees by sucrose stimulation of the antennae and/or the proboscis and can be rewarded by the delivery of food. The PER can be the object of both simple and complex forms of learning such as habituation and olfactory or tactile conditioning.

Habituation of the PER and Nicotinic Acetylcholine Receptors

Habituation of the PER is a simple form of learning in which the repetition of the gustatory stimulation of the antennae leads to a decrease in the probability of emitting the proboscis extension response. This test sheds light on the ability of an organism to modulate a reflex response. In *Aplysia*, a decrease in the synaptic strength has been advanced at the neuronal level between sensory and motor elements to explain the habituation of gill withdrawal reflex. The modification bears on homosynaptic depression linked to a decrease in the neurotransmitter released by the presynaptic terminal.[44] The habituation procedure has been successfully used in the honeybee to test its non-associative learning capabilities.[42,45]

We injected the nicotinic antagonists mecamylamine, α-BGT or MLA into the brain through the median ocellus, 15 minutes before a habituation session of the PER. The experiment consisted in repeatedly touching both antennae of starved animals with a weakly concentrated sucrose solution (29 mM) until they ceased responding. The criterion for habituation was to observe in the animals 5 consecutive stimulations inducing no response. All three nicotinic antagonists induced a decrease in the number of trials needed for habituation (ref. 46 and unpublished observation) (Fig. 2), an effect also induced by the neonicotinoid insecticide imidacloprid.[31,32,47] Conversely, honeybees fed with eserine, a blocker of AChE, needed more trials

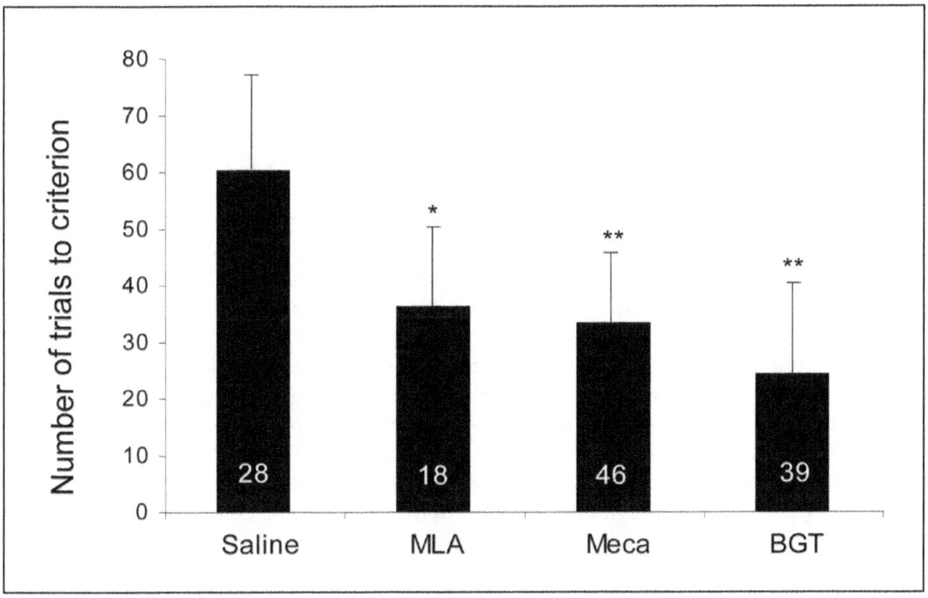

Figure 2. Effect of three nicotinic antagonists on habituation. Bars display the mean number of trials (with standard error) required to reach habituation criterion. Number of honeybees in each group is indicated in the bars. Student's *t* test (comparison to saline groups): *P < 0.05, **P < 0.01.

for habituation than controls.[42] The gustatory information from the antennae that induces the PER ends in the dorsal lobe of the deutocerebrum[18] and probably reaches the motor neurons located in the SOG driving the mouthpart movements. These primary gustatory afferents are supposed to be cholinergic, as α-BGT binding sites have been found in the dorsal lobes.[15,16] In addition, at least four nicotinic α subunits are expressed in the dorsal lobe.[48,49] This set of data suggests that nAChRs are present in the pathways involved in habituation. A similar facilitating effect on habituation was observed after injection of L-Name,[46] an inhibitor of NO-synthase enzyme that catalyzes the production of NO from L-arginine. This latter effect was unexpected, as previous results of the literature have reported that inhibiting NO-synthase increased the number of trials required to reach habituation.[50] Activation of NO-synthase following activation of nicotinic receptors sensitive to α-BGT is well established in the honeybee,[51,52] as in other insects[53] and leads to the production of NO. In insects as in mammals, NO acts as a signaling molecule in the nervous system.[54,55] One role that has been ascribed to NO in mammals is to participate in the experience-dependent plasticity of neuronal networks functioning. As a retrograde transmitter, NO could be responsible for long-term potentiation of glutamatergic synapses in the hippocampus, through increase of presynaptic neurotransmitter release.[56] We formulate the hypothesis that successive antennal sucrose stimulations reduce nicotinic neurotransmission between gustatory afferents and motor pathways. Blocking the nAChRs with the nicotinic antagonists, without blocking the PER, would enhance this process by mimicking the decrease in neurotransmission. The resulting decrease in NO production should lead to a decrease in ACh release from upstream pathways and consequently to a facilitation of habituation. By inhibiting the NO synthase, L-Name would contribute to diminishing the release of ACh and would facilitate habituation of the PER.

Classical Conditioning of the PER and Nicotinic Acetylcholine Receptors

The olfactory conditioning of the PER leads to the elicitation of proboscis extension by odor stimulation of the antennae. This behavioral paradigm is one of the paradigms most widely used in the honeybee to study the neurobiological bases of memory processes. It can be compared to olfactory shock-avoidance learning in *Drosophila* for the insights it has provided into the neurobiological substrates of memory processes.[57] The olfactory conditioning of the PER has been amply described elsewhere.[58] In this associative appetitive learning, restrained honeybees learn to associate an unknown odorant with a food reward. The behavioral response is the proboscis extension elicited by touching the antennae with a drop of sucrose solution. One forward paired presentation of odor and sucrose solution is generally sufficient to induce a conditioned response later on, when the odor is presented alone. In this Pavlovian conditioning, the unconditioned stimulus (US) leading to the proboscis extension is the sucrose solution, the conditioned stimulus (CS) is the odor and the conditioned response (CR) is the PER elicited by the odor. The most commonly used US in olfactory experiments is a compound US: antennal stimulation with sucrose is used to elicit the proboscis extension; then the bee that emits the proboscis extension response is fed with the sucrose solution. Feeding the bee for emitting the proboscis extension introduces an operant component in the conditioning procedure. To specify the nature of the association (operant or classical), an elegant omission procedure has been developed, in which the US was represented solely by the feeding component and was given each time the honeybee did not respond to the odorant. A high conditioning level was observed in these conditions where no link could develop between the conditioned response and the reinforcement, suggesting that the associative link is probably built between the odor and the sucrose solution and not between the response and the sucrose solution.[59]

The PER can also be conditioned to tactile cues, following the same rules of classical conditioning as for olfactory conditioning. In laboratory conditions, harnessed honeybees can associate a tactile antennal stimulus with a sucrose reinforcement delivered to the proboscis. Different protocols of mechanosensory conditioning have been developed. In the operant conditioning procedure, the bee is rewarded when its frequency of antennal contact with an object exceeds a certain threshold.[60] A classical procedure has been developed in which tactile stimulation of the base of the antennae is followed by the sucrose reward to the proboscis, irrespective of the antennal movements.[61] The protocol we used in the laboratory was modeled on the one developed by Erber et al[62]: bees were rewarded after scanning the surface of an object with their antennae in order to learn its texture properties; this form of conditioning follows both operant and classical rules. To clearly separate the tactile pathways for the conditioned stimulus from the reward pathway, the sugar solution was never delivered to the antennae. The sucrose was directly applied to the proboscis, to induce its extension and to reinforce the bee.[63]

When the honeybee has learned that an odor or a metal plate can predict the occurrence of food, the new information can be memorized for a short time or a long time. The duration of memory depends, in part, on the strength of the acquisition phase (i.e., the number of training trials). A single trial conditioning leads to a median-term memory (MTM) that lasts for several hours (until to 24 hours) whereas multiple trial conditioning induces long term memory (LTM) that underlies the retrieval performance from 24 hours and up.[64] An extensive study of honeybee LTM formed after massed and spaced learning has shown that the longer the inter-trial interval (10 min), the higher the retention rate. Protein synthesis inhibition during acquisition blocked the formation of late LTM (lLTM) (72 and 96 hours) after spaced and massed learning and blocked early LTM (eLTM) (24 and 48 hours) after spaced learning only.[65] Although in our experiments we used short inter-trial intervals (1 minute) for multiple-trial learning, in this way defining massed learning conditions, we found a significant decrease in eLTM (48 hours) and lLTM (72 hours) induced by protein synthesis inhibition (unpublished observation).

Our initial approach to studying the involvement of the cholinergic system in learning processes in the honeybee was to quantify the activity of AChE in groups of conditioned honeybees differing in their learning performance. We observed a decrease in AChE activity following a five-trial learning session in animals presenting a high performance level and not in animals presenting poor performance.[66] This result was in accordance with the work of Shapira et al[67] showing a down regulation of *AChE* gene expression in the brain of foraging bees presenting good learning capacities compared to nurses. However, in a second experiment performed on flies and bees we failed to find a correlation between the learning performance and the AChE rate individually in either *Drosophila* or in honeybees.[68] We pursued our neuropharmacological experiments by injecting nicotinic ligands into the brain to study the role of the nAChRs in memory formation and to specify the location of cholinergic brain structures involved in acquisition and retrieval processes.

Nicotinic Acetylcholine Receptors are Involved in Acquisition and Retrieval Processes

Using one-trial learning in the honeybee, we developed a pharmacological procedure consisting in brain injections of nicotinic ligands to study the role of the nAChRs in the processes of acquisition and in short-term memory. In a previous study centered on the role of muscarinic receptors in olfactory learning, we defined the relevance of using the one-trial learning procedure to clearly dissociate the different phases leading to the formation of memory.[40] Previous results obtained by Mercer and Menzel[69] have described the evolution of memory performance in retention tests performed in independent groups at different time intervals after one-trial learning sessions of olfactory conditioning. The retention curve established over the 2 hours following the conditioning trial was U-shaped, with an immediate high rate of CR and a minima during the first 5 minutes after learning. A high, stable performance level was reached for at least one hour, twenty minutes after conditioning. This phenomenon was first demonstrated by Kamin[70] on passive avoidance response in rodents. The "Kamin effect" has been subsequently found in

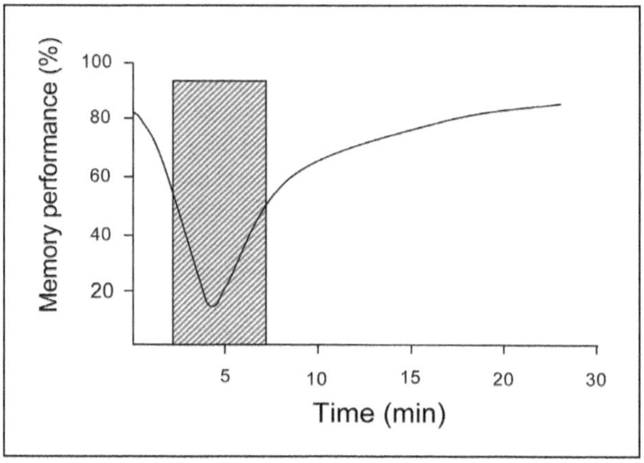

Figure 3. Theoretical time-course of the memory trace after one-trial conditioning in the honeybee. The hatched area indicates the phase of transition from short-term to long-term memory, the so-called consolidation process leading to the formation of a stable memory trace.

all animal species in which memory investigations have been conducted, indicating that a fundamental mechanism for memory formation is shared by several organisms (Fig. 3). The early curve trough is attributed to the consolidation processes that are triggered by acquisition and that will lead to the formation of a late stable memory. In the case of one trial conditioning, this period of information processing lasts several minutes in insects. Our experimental procedure was designed with these data in mind (Fig. 4). We injected mecamylamine, which appears in the vertebrate literature as the most potent nicotinic antagonist at the CNS level. Mecamylamine is also the molecule most commonly used in the experimental study of cognitive functions in rodents to avoid peripheral blocking effects of nicotinic antagonists. The drug was injected into the head hemolymph that ensures an overall diffusion of the solution into the brain. Honeybees were conditioned to olfactory or tactile learning. In a first series of experiments, mecamylamine

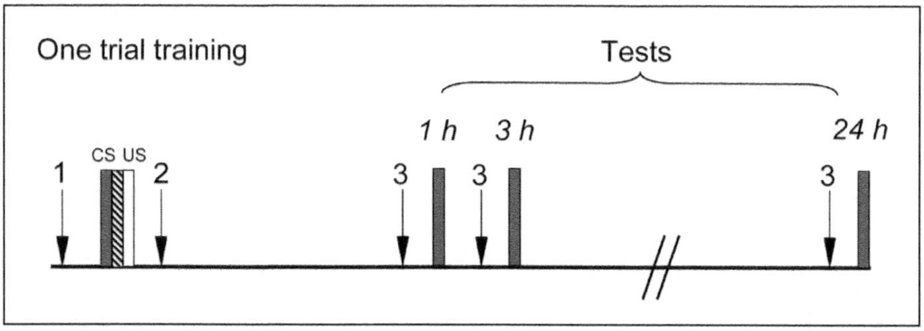

Figure 4. Design of the experimental protocol used to study the effect of the nicotinic antagonists on the different memory steps after one-trial conditioning (CS: conditioned stimulus, US: unconditioned stimulus). The hatched area indicates a 3-s overlap of CS and US. Retrieval tests evaluate short-term memory (1 hour), medium-term memory (3 hours) or long-term memory (24 hours). 1: The nicotinic antagonist is injected before training and may have an effect on the acquisition, the consolidation or on the retrieval processes. 2: The injection is performed immediately after training and may affect the consolidation and/or the retrieval processes. 3: The injection is performed after the consolidation phase and before one of the retrieval tests.

was injected 20 minutes after the conditioning trial in order to test the effect on retrieval processes. The drug induced an impairment of retention performance in the hour following the injection (Fig. 5A), with a recovery of the response rate at 80 minutes.[63,71] This indicates that mecamylamine was no longer active on retrieval processes at this time and that the effect of the drug lasted at least one hour. In supplementary tactile learning experiments, mecamylamine was injected 10 minutes before a retrieval test performed 3 or 24 hours after one-trial learning (see Fig. 4). In each case the drug blocked the recall of memory, indicating an effect on retrieval processes irrespective of the time of testing.[63] Mecamylamine injection was then performed 5

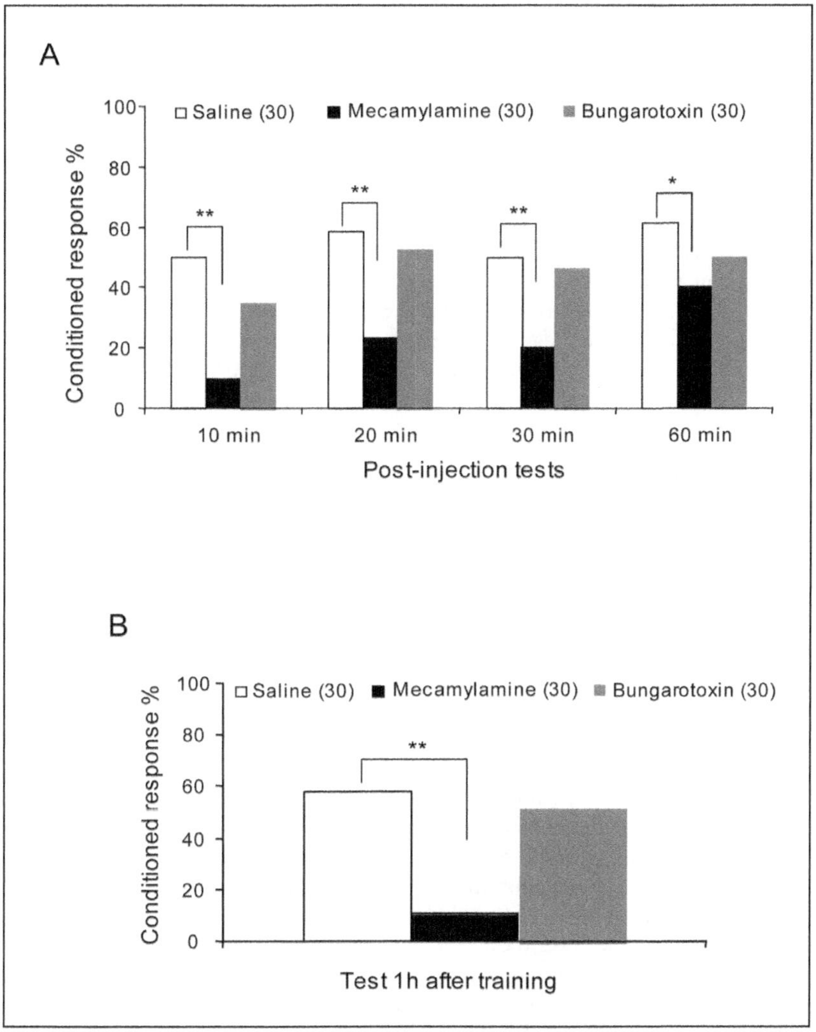

Figure 5. Effects of mecamylamine and α-bungarotoxin on one-trial olfactory learning. A) Effects on retrieval processes. The drugs were injected 20 min after training. The conditioned response (CR) was tested at four different times after injection in independent groups. B) Effect on acquisition processes. The drugs were injected 20 min before training and the honeybees were tested 1 hour after training. The numbers of honeybees in each group are presented in brackets. Fisher X^2 test (comparison to saline group): $^{**}P < 0.01$, $^{***}P < 0.001$.

minutes after the conditioning trial and no effect was found on retention in tests performed from 80 minutes to 24 hours after injection. Mecamylamine injected 20 minutes before training induced anterograde amnesia in animals tested 1 hour after training (Fig. 5B)[71] and at 3 and 24 hours.[63] Complementary experiments for olfactory learning were conducted to assess olfactory responsiveness in mecamylamine-treated animals. No difference was found between controls and treated bees in the spontaneous choice reaction to two odorants in an olfactometer.[71] Thus, an impairment of olfactory perception could be ruled out to account for the response decrease following mecamylamine injection. As mecamylamine injected immediately after learning had no effect on the consolidation process, we concluded that pretrial injection had no effect either on this memory process. Consequently, we concluded that pretrial injection of mecamylamine impaired acquisition by preventing the creation of the associative link between the CS and the US. These experiments clearly demonstrated an effect of mecamylamine on retrieval of olfactory[71] or tactile[63] information. Actually α-BGT was the first antagonist that we tested on one-trial olfactory or tactile learning. The negative results that we observed on memory (Fig. 5)[63,72] led us to use mecamylamine for further experiments. On the whole, our results suggest that different subtypes of nAChRs are probably present in the honeybee brain and they indicate that only the nicotinic receptor subtype sensitive to mecamylamine is activated by the single conditioning trial. They also show that blockade of nAChRs immediately after learning, during the phase of information processing leading to a stable memory trace, has no effect on the formation of memory. The hypotheses could be advanced that either the nAChRs that are blocked by mecamylamine are not necessary for the consolidation process, or the consolidation process following one-trial learning is so short that the injection performed 5 minutes posttrial comes too late to have an effect. Lastly, our results indicate that nAChRs are involved in acquisition and retrieval of olfactory and tactile information, irrespective of the sensory modality of the CS.

Brain Localization of Nicotinic Acetylcholine Receptors Involved in Acquisition and Retrieval Processes

The neuronal networks and structures involved in the successive steps of memory formation have been investigated with invasive brain experiments in the honeybee. Electrophysiological recordings have shown that individual neurons can modify their discharge activity related to the conditioning procedure, giving some information about learning at the cellular level. For example, the VUMx1 neuron, located in the suboesophageal ganglia, mediates the US and responds to the CS after pairing of the olfactory and gustatory stimulations.[73] This neuron belongs to a group of octopamine (OA)-immunoreactive neurons and innervates the AL glomeruli, the calyces of MB and the lateral protocerebrum, the main structures for odor processing. Pairing odor (CS) with OA injections into the ALs and MBs led to enhancement of PER and gave evidence for a role of OA in reinforcement processing during olfactory conditioning.[74] In the protocerebrum, a pedunculus extrinsic neuron, the PE1 neuron, shows a characteristic odor response before the training procedure. The discharge pattern is transiently modified after conditioning, suggesting an involvement of this neuron in short-term acquisition.[75,76] Inactivation by cooling of ALs and MBs at different times after one-trial olfactory learning showed a sequential activation of brain structures related to memory formation. Training triggered activation of ALs first, followed by activation of MBs.[77] Local anesthetics in α-lobes of MBs showed that MB activity is required for acquisition of complex olfactory tasks, but not of simple ones.[78]

We investigated the role of ALs in olfactory memory by injecting nicotine into the ALs at different concentrations before or after one-trial learning session. Nicotine (10^{-5}M) injected 20-min posttraining improved the PER response rate during tests performed 30 min and 60 min after injection. Other concentrations and other injection times had no effect on acquisition and retention rates. These results suggested that nicotine injected after the consolidation phase facilitates the retrieval processes needed to recall the learned information.[79] This result is consistent with the blockade of retrieval processes induced by the nicotinic antagonist mecamylamine.[71] The role of ALs in learning processes was also demonstrated by local injection of OA to replace the US, paired with olfactory stimulation of the antennae. This conditioning procedure produced paired-specific olfactory learning in the honeybees. The results suggested that the ALs are one of the brain sites where association between odor and reinforcement takes place and that they contribute to memory consolidation.[74] On the whole, ALs of the honeybee appear to be involved in all the information processing leading to the formation or the readout of memory.

We also investigated the role of the different parts of the MBs in olfactory memory by injecting mecamylamine 10 minutes before or 20 minutes after one-trial learning. Mecamylamine was injected either into the calyces (between the two calyces of each MB), which constitute the input of multisensory information, or into the α-lobes of the peduncles formed by the KC output. The results showed that pretrial injection of mecamylamine into calyces definitely suppresses the learning capabilities of the honeybee (Fig. 6A). Conversely an impairment of retrieval

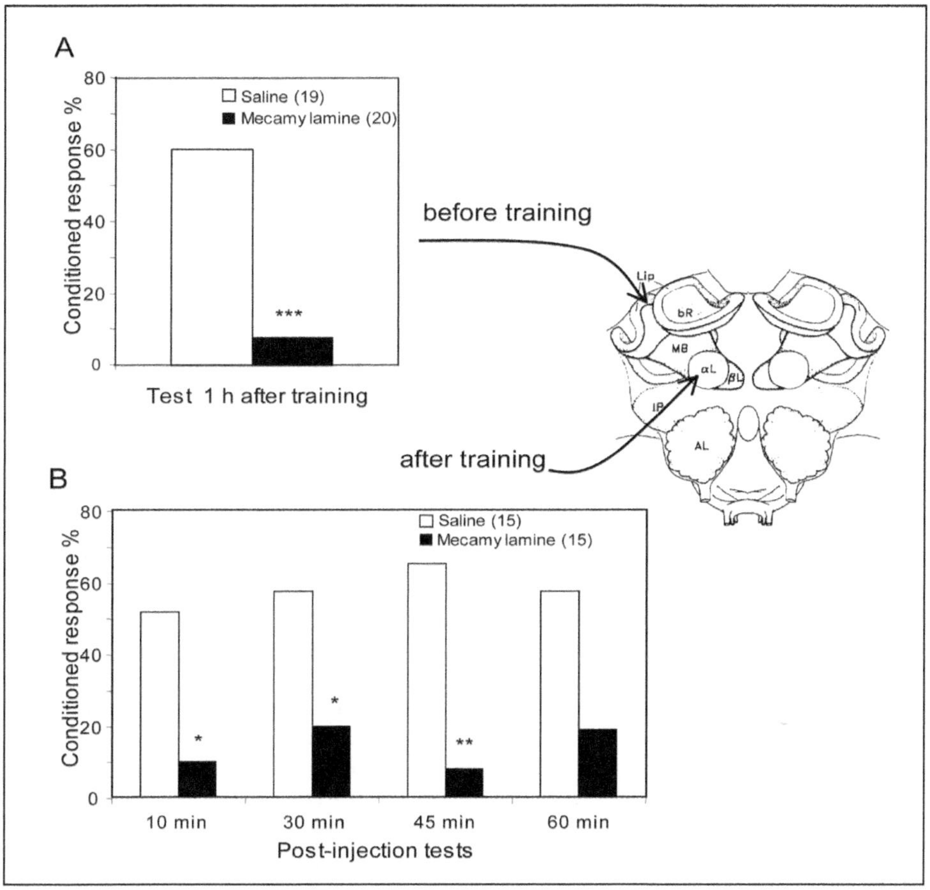

Figure 6. Effects of localized brain injection of mecamylamine on one-trial olfactory learning. A) Effect of mecamylamine injected 10 min before training between the median and the lateral calyces of each mushroom body. The honeybees were tested 1 hour after training. B) Effect of mecamylamine injected into the α lobe 20 min after training. The CR was tested at different times after injection in independent groups. The numbers of honeybees in each group are presented in brackets. Fisher X^2 test: *P < 0.05, **P < 0.01, ***P < 0.001.

processes was observed after posttrial mecamylamine injection into the α-lobe peduncles (Fig. 6B) but not after posttrial injection into the MB calyces.[80] Our experiments using local brain injections of mecamylamine demonstrate that nAChRs of the MBs are essential for associative learning processes that take place in the calyces of MBs. They also show that the recall of memory requires that the α-lobe nicotinic networks be intact. These experiments allow us to establish a functional map of the insect brain related to the formation and readout of memories. The MB calyces appear to be essential for the establishment of the full memory trace. During retrieval, the MB calyces seem to be no longer necessary. These results are in accordance with the work of Dubnau et al[81] and McGuire et al[82] in a mutant *Drosophila*. The *Shibire^{ts}* mutation was expressed in the fly MBs to reversibly disrupt synaptic transmission. Inactivation of MB signaling through output neurons during different phases of memory processing revealed a requirement for MB output signaling during memory retrieval but not during acquisition or consolidation. All these data suggest that the consolidated olfactory memory is localized in the α-lobes of MBs or alternatively, that access to a consolidated memory, whatever its location in the brain, requires the activation of α-lobe networks. In the honeybee, projection neurons from the antennal lobes to MB calyces form two tracts, the median antennoglomerular tract (mATG), which is putatively cholinergic[15] and the lateral ATG (lATG). A third tract, the medio-lateral ATG (mlAGT) sends olfactory projections from the ALs to the lateral protocerebrum (see Fig. 1). The absence of effect of mecamylamine injected into the calyces on retrieval could mean that either the calyces are no longer active during retrieval or that the lATG and the mlATG are sufficient to code the olfactory stimulus in such a way that they can activate the retrieval circuitry. These results reinforce a functional map of the olfactory brain with input regions (the ALs and the MB calyces) necessary to the associative processing of CS and US and an output region (MB α-lobes) involved in retrieval processes. The basics of this scheme are a distributed memory trace within the brain with functional properties that differ depending on their locations.

Alpha-BGT-Sensitive Nicotinic Acetylcholine Receptors and Long-Term Memory

The previous experiments, based on one-trial learning, investigated the role of the cholinergic system in short or medium-term memory. To study the role of the cholinergic system in long-term memory, multiple-trial olfactory or tactile learning was used. The learning session comprised 3 (olfactory learning) or 5 (tactile learning) trials, with one-minute inter-trial-intervals. We studied the involvement of nAChRs in the establishment of LTM by injecting the nicotinic antagonists mecamylamine, α-BGT or methyllycaconitine (MLA) into the brain, through the median ocellus. Pre-training injections of mecamylamine induced lower performance during conditioning but the retention level evaluated at medium-term (3-hour) and long-term (24-hour) intervals was equivalent between treated and control animals (Fig. 7A).[63,83] The effects of mecamylamine were then interpreted as a blockade of retrieval processes during learning, sufficient to explain the decrease in performance during the training session. In fact, from the second conditioning trial on, the odorant or tactile stimulus used as CS activates retrieval processes needed to elicit the conditioned response. A decrease in the conditioning rate during learning may reflect the bee's inability to retrieve the conditioned response but the formation of the memory is not impaired, as attested by the high conditioning rate of mecamylamine-treated animals, equivalent to that of controls, when tested for MTM and LTM. By contrast, α-BGT impaired the formation of long-term memory (24 hours) induced by multiple-trial learning (Fig. 7A), an effect also induced by MLA injection.[83] Alpha-BGT had no effect on medium-term memory (1 and 3 hours) and on retrieval processes.[63,83] Post-training injections of α-BGT (or MLA) decreased memory performance at 24 hours, whereas mecamylamine had no effect on long-term memory (Fig. 7B). These experiments showed that α-BGT (and MLA) blocked LTM formation while sparing MTM. Mecamylamine was effective in blocking retrieval processes, as also shown in one-trial learning experiments and had no effect on LTM. To explain this pharmacological dissociation,

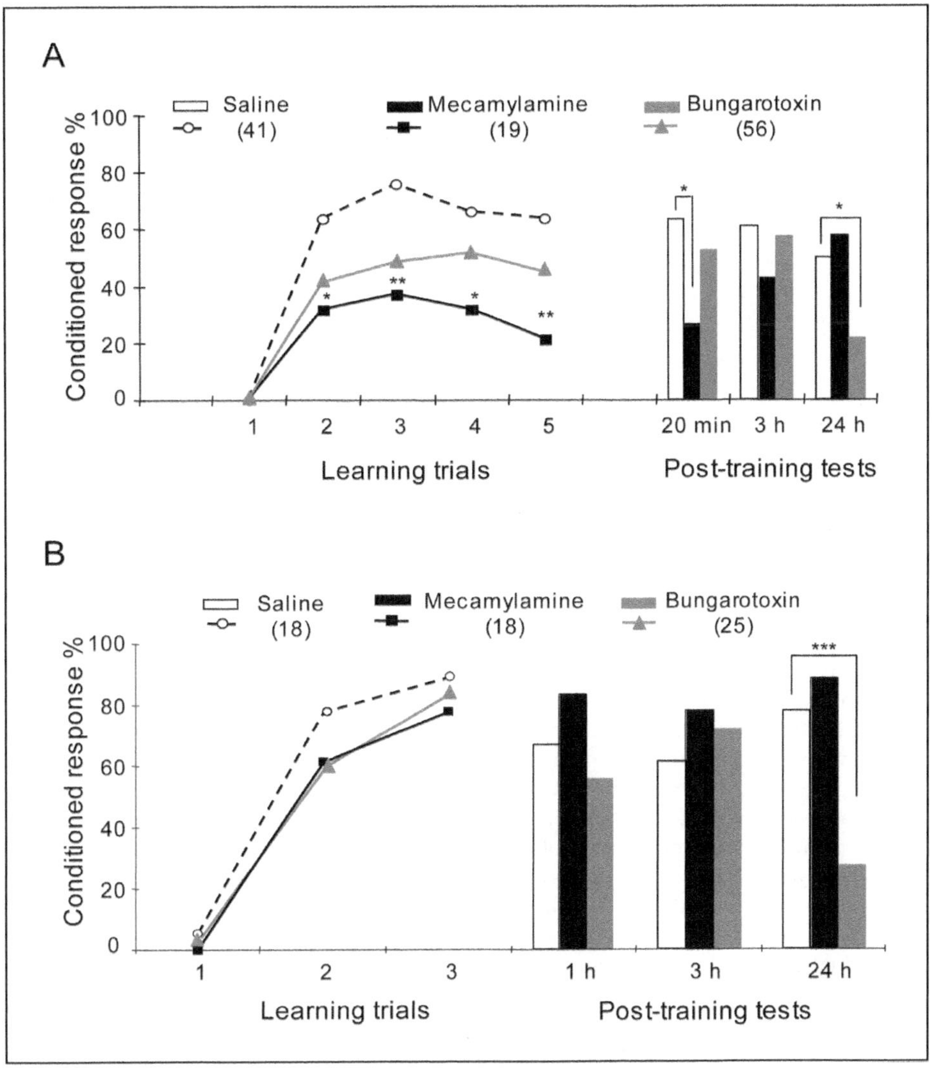

Figure 7. Effects of mecamylamine and α-bungarotoxin on multiple-trial learning. A) The drugs were injected 10 min before tactile learning. Each honeybee was tested at the three times after training. B) The drugs were injected 20 min after olfactory learning. Each honeybee was tested at the three times after training. The numbers of honeybees in each group are presented in brackets. Fisher X^2 test (comparison to saline group): $^*P < 0.05$, $^{***}P < 0.001$.

we advanced the hypothesis that retrieval processes involve the activation of α-BGT-insensitive nicotinic receptors and that LTM formation is dependent upon activation of α-BGT-sensitive nicotinic receptors. The existence of two nicotinic receptor subtypes, one α-BGT-sensitive and the other α-BGT-insensitive, is already well documented in insects[84-87] but our work is the first to postulate a function for these receptors in cellular plasticity supporting learning and memory. Our working hypothesis postulates that learning processes activate brain cholinergic pathways and ACh released from the presynaptic terminal may bind to all types of nicotinic receptors

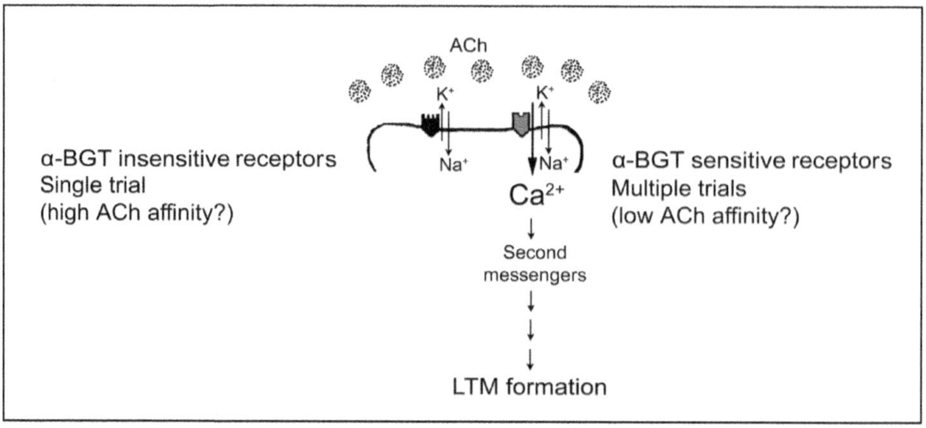

Figure 8. Schematic diagram illustrating the working hypothesis. The α-BGT-sensitive nAChRs (grey receptor on the right) require several trials to be activated by ACh and possess high permeability to calcium ions. The intracellular calcium increase triggers calcium-dependent cellular events linked to LTM formation. Alpha-BGT-insensitive nAChRs (black receptor on the left) may be activated during single-trial learning and retrieval test and during multiple-trial learning (each time ACh is released). Reprinted from Gauthier M et al. Neurobiol Learn Mem 2006; 86:164-174;[83] ©2006 with permission from Elsevier.

present (as well as to muscarinic receptors, a question that will be addressed in another chapter). One stimulation of these pathways like the ones induced by one-trial learning or CS presentation during the retrieval test will activate α-BGT-insensitive receptors. Multiple-trial learning will induce repetitive cholinergic stimulations of nAChRs and these multiple stimulations will trigger activation of the α-BGT-sensitive receptors. This in turn, will trigger intracellular events involved in LTM formation (Fig. 8). Understanding how α-BGT-sensitive nAChRs are regulated and how they detect multiple stimulations will be a major challenge for our future research. As has been shown in studies on neuronal nAChRs of vertebrates, a subpopulation of honeybee α-BGT-sensitive nAChRs has been described to trigger, upon activation, an influx of Ca^{2+} ions into the cell,[24,51,52] a phenomenon also described in crickets[88] and *Drosophila*.[89-91] Interestingly, the work of Campusano et al[91] shows that repetitive exposure of Kenyon cells to nicotine results in a calcium-dependent plasticity of the nAChR-mediated response through cAMP signaling. Intracellular Ca^{2+} may also act as a key second messenger for the triggering of intra-cellular cascades leading to LTM. In insects, the enzyme NO synthase is activated by the Ca^{2+} ions[92,93] and NO is specifically involved in LTM formation.[46,94] Since in vitro experiments have shown that α-BGT-sensitive nAChRs can trigger NO synthesis in insects,[52,53] α-BGT- and MLA-sensitive nAChRs may specifically trigger the NO release involved in LTM formation during multiple-trial learning. Our data suggest that activation of nAChRs can induce cellular plasticity that could contribute to information processing supporting LTM formation.

Conclusion and Outlook: Using RNA Interference to Create Reversible Mutant Honeybees for Memory

The differences between the nicotinic receptors' properties are linked to their subunit composition. Four *A. mellifera* nAChR subunits have been cloned in the lab and were shown to be expressed in the honeybee brain.[48,49,95] With the sequencing of the honeybee genome,[96] the complete *A. mellifera* nAChR gene family has now been described.[27] Nine α and two β subunits have been identified and each of these subunits can potentially combine with each another to form different subtypes of homomeric or heteromeric nAChRs. Moreover, the alternative splicing of some of these subunits such as Amelα4 or Amelα6 increases the complexity of nAChRs. Individual subunits can confer distinct pharmacological properties on a receptor such as its sensitivity to insecticides[97] or can determine specific regulation of those nAChR receptors whose activation depends on repeated stimulation and triggers long-term memory formation.[83] It has been shown in cockroaches that nAChR properties can be modulated by phosphorylation/dephosphorylation mechanisms.[85] In *Apis mellifera*, different splice sites in Amelα3 give rise to two variants, which present different intracellular domains including different phosphorylation sites.[27] Since phosphorylation of the intracellular loop is involved in regulating several aspects of receptor function such as desensitization, drug sensitivity and aggregation, the two splice variants have the potential to alter several receptor properties. Amelα7 mRNA was detected by in situ hybridization in antennal lobe and dorsal lobe neurons and Kenyon cells of MB. The sequence of the Amelα7 subunit exhibits high homologies with the α7 subunit vertebrate sequence.[49] Interestingly, α-BGT and MLA are the most selective antagonists of the α7 homomeric neuronal nAChRs in vertebrates and these receptors have low affinity to ACh. A detailed knowledge of the composition and pharmacology of the honeybee nAChRs is necessary to fully understand the effects of the nicotinic ligands on behavior. Specific α or β subunit deletion is now available in the honeybee using the RNA interference (RNAi) technique. Introducing double strand RNA (dsRNA) into cells results in posttranscriptional silencing of target genes through RNAi. This technique has been used successfully in the honeybee to study the role of the octopaminergic reinforcing pathway in olfactory learning.[98] Possessing the genome sequence for the 11 honeybee nAChR subunits potentially allows us to construct dsRNA against each subunit and test the corresponding gene silencing on behavioral functions. Using a combination of dsRNA injection and behavior, we will test the hypothesis that, just as in vertebrates, Amel α7 subunit would form homomeric receptors and that this nAChR subtype should be involved in LTM formation.

Acknowledgements

The author is grateful to V. Raymond-Delpech for her helpful comments on the manuscript and to P. Haas-Hammel for proofreading.

References

1. Dani JA, Bertand D. Nicotinic acetylcholine receptors and nicotinic cholinergic mechanisms of the central nervous system. Ann RevPharmacol Toxicol 2007; 47:699-729.
2. Gotti C, Clementi F. Neuronal nicotinic receptors: from structure to pathology. Progress Neurobiol 2004; 74:363-396.
3. Hasselmo ME. The role of acetylcholine in learning and memory. Curr Opinion Neurobiol 2006; 16:710-715.
4. Jensen AA, Frølund B, Liljrfors T et al. Neuronal nicotinic receptors: structural revelations, target identification and therapeutic inspirations. J Med Chem 2005; 48:4705-4745.
5. Dani JA. Nicotine mechanisms in Alzheimer's disease. Overview of nicotinic receptors and their roles in the central nervous system. Biol Psychiatry 2001; 49:166-174.
6. Breer H. Neurochemical aspects of cholinergic synapes in the insect brain. In: Gupta AP ed. Arthropod brain. Its evolution, development, structure and functions. New-York: Wiley, 1987:415-437.
7. Oleskevitch S. Cholinergic synaptic transmission in insect mushroom bodies in vitro. J Neurophysiol 1999; 82:1091-1096.
8. Lee D, O'Dowd DK. Fast excitatory synaptic transmission mediated by nicotinic acetylcholine receptors in Drosophila neurons. J Neurosci 1999; 19:5311-5321.
9. Heinrich R, Hedwig B, Elsner N. Cholinergic activation of stridulatory behavior in the grasshopper Omocestus viridulus (L.). Exp Biol 1997; 200:1327-1337.
10. Gauglitz S, Pflüger HJ. Cholinergic transmission via central synapses in the locust nervous system. J Comp Physiol A 2001; 187:825-836.

11. Breer H, Sattelle DB. Molecular properties and functions of insect acetylcholine receptors. J Insect Physiol 1987; 33:771-790.
12. Bicker G. Histochemistry of classical neurotransmitters in antennal lobes and mushroom bodies of the honeybee. Micros Res Tech 1999; 45:174-183.
13. Gorczyca M, Hall JC. Identification of a cholinergic synapse in the giant fiber pathway of Drosophila using conditional mutations of acetylcholine synthesis. J Neurogenet 1984; 1:289-313.
14. Homberg U. Neurotransmitters and neuropeptides in the brain of the locust. Micros Res Tech 2002; 56:189-209.
15. Kreissl S, Bicker G. Histochemistry of acetylcholinesterase and immunocytochemistry of an acetylcholine receptor-like antigen in the brain of the honeybee. J Comp Neurol 1989; 286:71-84.
16. Scheidler A, Kaulen P, Bruning G et al. Quantitative autoradiographic localization of [^{125}I] α-bungarotoxin binding sites in the honeybee brain. Brain Res 1990; 534:332-335.
17. Yusuyama K, Meinertzhagen IA, Shürmann FW. Synaptic organization of the mushroom body calyx in Drosophila melanogaster. J Comp Neurol 2002; 445:211-226.
18. Haupt S. Central gustatory projections and side-specificity of operant antennal muscle conditioning in the honeybee (Apis mellifera L.). J Comp Physiol A 2007; 193:523-535.
19. Osborne RH. Insect neurotransmission: neurotransmitters and their receptors. Pharmacol Ther 1996; 69:117-142.
20. Gundelfinger ED, Schulz R. Insect nicotinic acetylcholine receptors: gene, structure, physiological and pharmacological properties. In: Clementi F, Fornassari D, Gotti C eds. Neuronal nicotinic receptors. Berlin, Heidelberg, New-York: Springer, 2000:497-521.
21. Barbara G, Zube C, Rybak J et al. Acetylcholine, GABA and glutamate induce ionic currents in cultured antennal lobe neurons of the honeybee, Apis mellifera. J Comp Physiol A 2005; 191:823-836.
22. Barbara GS, Grünewald B, Paute S et al. Cholinergic ionic currents of cultured antennal lobe neurons from adult honeybee brains. Invert Neurosci 2008; 8:19-29.
23. Déglise P, Grünewald B, Gauthier M. The insecticide imidacloprid is a partial agonist of the honeybee neuronal nicotinic receptor. Neurosci Lett 2002; 321:13-16.
24. Goldberg F, Grünewald B, Rosenboom H et al. Nicotinic acetylcholine currents of cultured Kenyon cells from the mushroom bodies of the honey bee Apis mellifera. J Physiol London 1999; 514:759-768.
25. Wüstenberg DG, Grünewald B. Pharmacology of the neuronal nicotinic acetylcholine receptor of cultured Kenyon cells of the honeybee, Apis mellifera. J Comp Physiol A 2004; 190:807-821.
26. Jones AK, Brown LA, Sattelle DB. Insect nicotinic acetylcholine receptors: from genetic model organism to vector, pest and beneficial species. Invert Neurosci 2007; 7:67-73.
27. Jones AK, Raymond-Delpech V, Thany SH et al. The nicotinic acetylcholine receptor gene family of the honeybee, Apis mellifera. Genome Res 2006; 16:1422-1430.
28. Millar N, Denholm I. Nicotinic acetylcholine receptors: targets for commercially important insecticides. Invert Neurosci 2007; 7:53-66.
29. Tomizawa M, Casida JE. Neonicotinoid insecticide toxicology. Annu Rev Pharmacol Toxicol 2005; 45:247-268.
30. Wenzel B, Hedwig B. Neurochemical control of cricket stridulation revealed by pharmacological microinjections into the brain. J Exp Biol 1999; 202:2203-2216.
31. Lambin M, Armengaud C, Raymond S et al. Imidacloprid-induced facilitation of the proboscis extension reflex habituation in the honeybee. Arch Insect Biochem Physiol 2001; 48:129-134.
32. Armengaud C, Lambin M, Gauthier M. Effects of imidacloprid on the neural processes of memory in the honeybee. In: Devillers J, Pham-Délègue MH, eds. Honeybees: Estimating the environmental impact of chemicals. London: Taylor and Francis, 2002:85-100.
33. El Hassani AK, Dacher M, Gary V et al. Effects of sublethal doses of acetamiprid and thiamethoxam on the behavior of the honeybee (Apis mellifera). Arch Environm Contam Toxicol 2008; 54:653-661.
34. Shang Y, Claridge-Chang A, Sjulson L et al. Excitatory local circuits and their implications for olfactory processing in the fly antennal lobe. Cell 2007; 128:601-612.
35. Olsen S, Bhandawat V, Wilson R. Excitatory interactions between olfactory processing channels in the Drosophila antennal lobe. Neuron 2007; 54:89-103.
36. Bainton RJ, Tsai LT-Y, Singh C et al. Dopamine modulates acute responses to cocaine, nicotine and ethanol in Drosophila. Curr Biol 2000; 10:187-194.
37. Hou J, Kuromi H, Fukasawa Y et al. Repetitive exposure to nicotine induces hyper-responsiveness via the camp/PKA/Creb signal pathway in Drosophila. J Neurobiol 2004; 60:249-261.

38. Ismail N, Robinson GE, Fahrbach SE. Stimulation of muscarinic receptors mimics experience-dependent plasticity in the honeybee brain. PNAS 2006; 103:207-211.
39. Ismail N, Christine S, Robinson GE et al. Pilocarpine improves recognition of nestmates in young honeybees. Neurosci lett 2008; 439:178-181.
40. Gauthier M, Cano Lozano V, Zaoujal A et al. Effects of intracranial injections of scopolamine on olfactory conditioning retrieval in the honeybee. Behav Brain Res 1994; 63:145-149.
41. Cano Lozano V, Gauthier M. Effects of the muscarinic antagonists atropine and pirenzepine on olfactory conditioning in the honeybee. Pharmacol Biochem Behav 1998; 59:903-907.
42. Braun G, Bicker G. Habituation of an appetitive reflex in the honeybee. J Neurophysiol 1992; 67:588-598.
43. Giurfa M. Behavioral and neural analysis of associative learning in the honeybee: a taste from the magic well. J Comp Physiol A 2007; 193:801-824.
44. Castellucci VF, Pinsker H, Kupfermann I et al. Neuronal mechanisms of habituation and deshabituation withdrawal reflex in Aplysia. Science 1970; 167:1745-1748.
45. Bicker G, Hahnlein I. Long-term habituation of an appetitive reflex in the honeybee. Neuroreport 1994; 6:54-56.
46. Dacher M, Gauthier M. Involvement of NO-synthase and nicotinic receptors in learning in the honeybee. Physiol Behav 2008; 95:200-207.
47. Guez D, Suchail S, Gauthier M et al. Contrasting effects of imidacloprid on habituation in 7- and 8-day-old honeybees (Apis mellifera). Neurobiol Learn Mem 2001; 76:183-191.
48. Thany SH, Lenaers G, Crozatier M et al. Identification and localization of the nicotinic acetylcholine receptor alpha3 mRNA in the honey bee brain Apis mellifera. Insect Mol Biol 2003; 12:255-262.
49. Thany SH, Crozatier M, Raymond-Delpech V et al. Apisα2, Apisα7-1 and Apisα7-2 Three new neuronal nicotinic acetylcholine receptor subunits in the brain of the honeybee Apis mellifera. Gene 2005; 344:125-132.
50. Müller U, Hildebrandt H. Nitric oxide/cGMP-mediated protein kinase A activation in the antennal lobes plays an important role in appetitive reflex habituation in the honeybee. J Neurosci 2002; 22:8739-8747.
51. Bicker G, Kreissl S. Calcium imaging reveals nicotinic acetylcholine receptors on cultured mushroom body neurons. J Neurophysiol 1994; 71:808-810.
52. Bicker G. Transmitter-induced calcium signalling in cultured neurons of the insect brain. J Neurosci Methods 1996; 69:33-41.
53. Zayas RM, Qazi S, Morton DB et al. Nicotinic-acetylcholine receptors are functionally coupled to the nitric oxide/cGMP-pathway in insect neurons. J Neurochem 2002; 83:421-431.
54. Müller U. The nitric oxide system in insects. Progress Neurobiol 1997; 51:363-381.
55. Bicker G. Nitric oxide: an unconventional messenger in the nervous system of an orthopteroid insect. Arch Insect Biochem Physiol 2001; 48:100-110.
56. Schuman EM, Madison DV. Nitric oxide and synaptic function. Ann Rev Neurosci 1994; 17:153-183.
57. Tully T. Drosophila learning and memory revisited. Trends Neurosci 1987; 10:330-335.
58. Bitterman ME, Menzel R, Fietz A et al. Classical conditioning of proboscis extension in honeybees (Apis mellifera). J Comp Psychol 1983; 97:107-119.
59. Menzel R, Bitterman ME. Learning by honeybees in unnatural situation. In: Huber F, Markl H, eds. Neuroethology and behavioural physiology. Berlin Heidelberg: Springer Verlag, 1983; 206-215.
60. Kish J, Erber J. Operant conditioning of antennal movements in the honeybee. Behav Brain res 1999; 99:93-102.
61. Giurfa M, Malun D. Associative mechanosensory conditioning of the proboscis extension reflex in the honeybee. Learn Mem 2004; 11:294-302.
62. Erber J, Kierzek S, Sander E et al. Tactile learning in the honeybee. J Comp Physiol A 1998; 183:737-744.
63. Dacher M, Lagarrigue A, Gauthier M. Antennal tactile learning in the honeybee: effect of nicotinic antagonists on the memory dynamics. Neurosci 2005; 130:37-50.
64. Menzel R. Memory dynamics in the honeybee. J Comp Physiol A 1999; 85:323-340.
65. Menzel R, Manz G, Menzel R et al. Massed and spaced learning in honeybees: the role of CS, US, the intertrial interval and the test interval. Learn Mem 2001; 8:198-208.
66. Gauthier M, Belzunces LP, Zaoujal A et al. Modulatory effect of learning and memory on honeybee brain acetylcholinesterase activity. Comp Bioch Physiol 1992; 103C:91-95.

67. Shapira M, Thompson CK, Soreq H et al. Changes in neuronal acetylcholinesterase gene expression and division of labor in honeybee colonies. J Mol Neurosci 2001; 17:1-12.
68. Fresquet N, Fournier D, Gauthier M. A new attempt to assess the effects of learning abilities on the cholinergic system: Studies on fruitflies and honeybees. Comp Biochem Physiol 1998; 119B:349-353.
69. Mercer A, Menzel R. The effects of biogenic amines on conditioned and unconditioned responses to olfactory stimuli in the honeybee Apis mellifera. J Comp Physiol A 1982; 145:363-368.
70. Kamin LJ. The retention of an incompletely learned avoidance response. J Comp Physiol Psychol 1957; 50:457-460.
71. Cano Lozano V, Bonnard E, Gauthier M et al. Mecamylamine-induced impairment of acquisition and retrieval of olfactory conditioning in the honeybee. Behav Brain Res 1996; 81:215-222.
72. Cano Lozano V. Etude du rôle du système cholinergique dans les processus de mémorisation chez l'abeille. PhD thesis, Université Paul Sabatier Toulouse III, 1997.
73. Hammer M. An identified neuron mediates the unconditioned stimulus in associative olfactory learning in honeybees. Nature 1993; 366:59-63.
74. Hammer M, Menzel R. Multiple sites of associative odor learning as revealed by local brain microinjections of octopamine in honeybees. Learn Mem 1998; 5:146-156.
75. Mauelshagen J. Neural correlates of olfactory learning in an identified neuron in the honeybee brain. J Neurophysiol 1993; 69:609-625.
76. Okada R, Rybak J, Manz G et al. Learning-related plasticity in PE1 neuron and other mushroom body-extrinsic neurons in the honeybee brain. J Neurosci 2007; 27:11736-11747.
77. Erber J, Mashur TH, Menzel R. Localization of short-term memory in the brain of the bee, Apis mellifera. Physiol Entomol 1980; 5:343-358.
78. Devaud JM, Blunk A, Podufall J et al. Using local anaesthetics to block neuronal activity and map specific learning tasks to the mushroom bodies of an insect brain. Europ J Neurosci 2007; 26:3193-3206.
79. Thany SH, Gauthier M. Nicotine injected into the antennal lobes induces a rapid modulation of sucrose threshold and improves short-term memory in the honeybee Apis mellifera. Brain Res 2005; 1039:216-219.
80. Cano Lozano V, Armengaud C, Gauthier M. Memory impairment induced by cholinergic antagonists injected into the mushroom bodies of the honeybee. J Comp Physiol A 2001; 187:249-254.
81. Dubnau J, Grady L, Kitamoto T et al. Disruption of neurotransmission in Drosophila mushroom body blocks retrieval but not acquisition of memory. Nature 2001; 411:476-480.
82. McGuire S, Le PT, Davis RL. The role of Drosophila mushroom body signaling in olfactory memory. Science 2001; 293:1330-1333.
83. Gauthier M, Dacher M, Thany S et al. Involvement of alpha-bungarotoxin-sensitive nicotinic brain receptors in long-term memory formation in the honeybee (Apis mellifera). Neurobiol Learn Mem 2006; 86:164-174.
84. Buckingham S, Lapied B, Le Corronc H et al. Imidacloprid actions on insect neuronal acetylcholine receptors. J Exp Biol 1997; 200:2685-2692.
85. Courjaret R, Lapied B. Complex intracellular messenger pathways regulate one type of neuronal alpha-bungarotoxin-resistant nicotinic acetylcholine receptors expressed in insect neurosecretory cells (dorsal unpaired median neurons). Mol Pharmacol 2001; 60:80-91.
86. Lansdell SJ, Millar NS. The influence of nicotinic receptor subunit composition upon agonist, alpha-bungarotoxin and insecticide (imidacloprid) binding affinity. Neuropharmacology 2000; 39:671-679.
87. Tomizawa M, Casida JE. Structure and diversity of insect nicotinic acetylcholine receptors. Pest Manag Sci 2001; 57:914-922.
88. Cayre M, Buckingham SD, Yagodin S et al. Cultured insect mushroom body neurons express functional receptors for acetylcholine, GABA, glutamate, octopamine and dopamine. J Neurophysiol 1999; 81:1-14.
89. Yu D, Baird GS, Tsien RY et al. Detection of calcium transients in Drosophila mushroom body neurons with camgaroo reporters. J Neurosci 2003; 23:64-72.
90. Jiang SA, Campusano JM, Su H et al. Drosophila mushroom body Kenyon cells generate spontaneous calcium transients mediated by PLTX-sensitive calcium channels. J Neurophysiol 2005; 94:491-500.
91. Campusano JM, Su H, Jiang SA et al nAChR-mediated calcium responses and plasticity in Drosophila Kenyon cells. Develop Neurobiol 2007; 67:1520-1532.

92. Müller U. Ca^{2+}/calmodulin-dependent nitric oxide synthase in Apis mellifera and Drosophila melanogaster. Europ J Neurosci 1994; 6:1362-1370.
93. Müller U, Bicker G. Calcium-activated release of nitric oxide and cellular distribution of nitric oxide-synthesizing neurons in the nervous system of the locust. J Neurosci 1994; 14:7521-7528.
94. Müller U. Inhibition of nitric oxide synthase impairs a distinct form of long-term memory in the honeybee, Apis mellifera. Neuron 1996; 16:541-549.
95. Thany SH. Caractérisation des différentes sous-unités des récepteurs nicotiniques neuronaux chez l'abeille, Apis mellifera. PhD thesis, Université Paul Sabatier Toulouse III, 2004.
96. Honeybee Genome Sequencing Consortium. Insights into social insects from the genome of the honeybee Apis mellifera. Nature 2006; 443:931-949.
97. Romanelli MN, Gualtieri F. Cholinergic nicotinic receptors: competitive ligands, allosteric modulators and their potential applications. Med Res Rev 2003; 23:393-426.
98. Farooqui T, Robinson K, Vaessin H et al. Modulation of early olfactory processing by an octopaminergic reinforcement pathway in the honeybee. J Neurosci 2003; 23:5370-5380.

INDEX

L

M

N

P

R

S

T

V

X

Batch number: 10762179

Printed by Printforce, the Netherlands